PLASTIC FUTURES IN ARCHITECTURE

This volume details the recent efforts of developing sustainable approaches to reclaim synthetic polymers and explore natural polymers in design production. Not long ago, plastic was envisioned as a miraculous material. Now plastic pollution has become an inseparable part of the environment. How can the built environment make use of this prolific and 'miraculous' material in an environmentally responsible way?

Plastic Futures in Architecture focuses on plastic derivatives and complications in three main modes: Being, Becoming, and Beyond. "Being" presents the concerns and prospects of plastic continuation. "Becoming" provides insight into the technological advancements in the processing and production of plastic in the built environment. "Beyond" discusses the experimental transformation of plastic applications from synthetic polymers to more sustainable polymers in design. It focuses on projects that outline a journey from awareness to innovation, recycling synthetic polymers and exploring the use of bio-based natural polymers through rapidly developing digital crafting capabilities and techniques.

These project discussions will serve as an invaluable resource for students, educators, architects, product designers, practitioners in global sustainability, material innovation, digital fabrication technologies, and computational design.

Ehsan Baharlou, Dr.-Ing., is an Assistant Professor of Architecture and the Director of Computational Tectonics Lab at the School of Architecture, University of Virginia. His research agenda unfolds through three interconnected streams: Computationally Driven Material Ecologies, Circular Robotic Fabrication for Performance and Living, and Regenerative Architecture. He investigates regenerative architectural technologies across scales—from microscale material formulations to full-scale assemblies. His work has been internationally exhibited at the French Pavilion in Venice Biennale 2025 and at the 2025 Chicago Architecture Biennial. He has also co-edited a special issue of the *International Journal for Architectural Computing (IJAC)*, advancing computational design discourse by foregrounding ecological levers such as autonomy and interdependence.

PLASTIC FUTURES IN ARCHITECTURE

From Recycled Synthetic to Bio-Based Polymers

Edited by Ehsan Baharlou

Routledge
Taylor & Francis Group
LONDON AND NEW YORK

Designed cover image: Ehsan Baharlou

First published 2026
by Routledge
4 Park Square, Milton Park, Abingdon, Oxon OX14 4RN

and by Routledge
605 Third Avenue, New York, NY 10158

Routledge is an imprint of the Taylor & Francis Group, an informa business

British Library Cataloguing-in-Publication Data
A catalogue record for this book is available from the British Library

Library of Congress Cataloging-in-Publication Data
Names: Baharlou, Ehsan editor
Title: Plastic futures in architecture : from recycled synthetic to bio-based polymers / edited by Ehsan Baharlou.
Identifiers: LCCN 2026005890 | ISBN 9781041011880 hbk | ISBN 9781041011873 pbk | ISBN 9781003613589 ebk
Subjects: LCSH: Plastics in building | Biopolymers–Industrial applications | Sustainable architecture
Classification: LCC TA668 .P545 2026
LC record available at https://lccn.loc.gov/2026005890

ISBN: 9781041011880 (hbk)
ISBN: 9781041011873 (pbk)
ISBN: 9781003613589 (ebk)

DOI: 10.4324/9781003613589

Typeset in Times New Roman
by Newgen Publishing UK

CONTENTS

CONTRIBUTORS

Phoebe Crisman, AIA, is Professor of Architecture and Director of the interdisciplinary Global Studies program at the University of Virginia. She teaches design studios and architecture, urban, and sustainability theory. Her public-interest research and design practice with Crisman+ Petrus Architects intertwines sustainable and just strategies, ecological restoration, and material exploration.

Blaine Brownell is an architect and researcher of emergent materials and sustainable building technologies. He is Fellow of the American Institute of Architects (FAIA) and Professor and Director of the David R. Ravin School of Architecture at the University of North Carolina at Charlotte. A former Fulbright scholar to Japan, he has authored nine books on advanced and sustainable materials for architecture and design, including *Matter in the Floating World, Material Strategies,* and the four-volume Transmaterial series. He writes the Mind & Matter column for *Architect* magazine, and his work has been published in over 70 architecture, design, science, and news journals, including *The New York Times*, *The London Times*, *The Wall Street Journal*, and *Nature*. His most recent book, *The Pandemic Effect*, addresses the inoculation of the built environment.

Karola Dierichs is an architect and a scientist in the Department of Biomaterials at the Max Planck Institute of Colloids and Interfaces. In her research, she integrates the fields of materials design and minimal machines for architectural construction. Here, the main goal is to establish architecture as sourced from and embedded in a given environment.

Maria Paz Gutierrez is Professor at UC Berkeley, focused on materials invention. She holds a PhD in Building Science and Materials Engineering from the University of Cambridge and an M Arch from the University of Pennsylvania. Design and scientific awards include the US NSF, and her work has been featured in the Architecture Venice Biennale.

Philip F. Yuan is Professor and Dean of the College of Architecture and Urban Planning at Tongji University. He is also an honored FAIA and the Editor-in-Chief of the journal *Architectural*

Intelligence. He has received numerous prestigious awards both domestically and internationally, including the UIA International Auguste Perret Award for Architectural Technology (the first Chinese recipient), the AIA 2022 International Architecture Honor Award, the 2022 Dezeen Design Award, the 2024 National Teaching Innovation Competition for University Teachers (National Second Prize, Shanghai Special Prize), etc.

Jiapan Sun is a PhD candidate at the Shanghai Research Institute for Intelligent Autonomous Systems, Tongji University. He received his Master's degree from University College London. His research focuses on the development and application of solid waste and bio-based low-carbon building materials, with an emphasis on their integration and innovation in sustainable architecture and digital construction. His related work has been published in internationally renowned conferences such as *CAAD Futures*, *CAADRIA*, and *eCAADe*.

Liming Zhang is a PhD candidate in Architecture at Tongji University, specializing in architectural robotics and 3D printing research and development. Over the past six years, he has participated in several national and provincial key R&D projects related to construction robotics and 3D printing, and has applied for more than ten invention patents. As the technical lead for 3D printing, he has contributed to over ten demonstrative projects, including the Light of the Internet Expo Center for the World Internet Conference in Wuzhen, the World Artificial Intelligence Conference venue, and the outdoor pavilion of the China Pavilion at the Venice Architecture Biennale.

Kathy Velikov is Professor of Architecture and Associate Dean for Research and Creative Practice at the University of Michigan Taubman College of Architecture and Urban Planning. A founding partner of RVTR, her research advances environmentally, technologically, and socially focused built environments through material assembly prototyping and urban infrastructure design.

Wes McGee is Professor in Architecture, Director of the Fabrication and Robotics Lab at the University of Michigan Taubman College of Architecture and Urban Planning, and Principal at Matter Design. His research lies at the intersection of design, engineering, materials, and manufacturing processes as they relate to the built environment.

Daniel Tish is a Design Critic in Architecture at the Harvard University Graduate School of Design. His work lies at the intersection of digital fabrication, material science, sustainability, and computation. He holds a Doctor of Design from the Harvard GSD, has previously taught at the University of Michigan, and has exhibited work internationally.

Geoffrey Thün is Professor of Architecture at Taubman College of Architecture and Urban Planning, and a founding partner of RVTR. He currently serves as the Associate Vice President for Research: Social Sciences, Humanities and the Arts at University of Michigan.

Marshall Prado is Associate Professor of Design and Structural Technology at the University of Tennessee, Knoxville, College of Architecture and Design. His research explores integrated computational design, robotic fabrication, additive manufacturing, and lightweight fiber composite systems for architectural applications.

Julia Koerner-Al-Rawi is an award-winning Austrian designer working at the convergence of architecture, product, and fashion design, specialized in 3D printing. She is the founder of JK Design/JK3D and is a Professor at UCLA. Her work includes 3D-Printed Haute Couture and Academy Award-winning costumes for the Hollywood blockbuster *Black Panther*.

Virginia San Fratello is Professor and the Chair of the Department of Design at SJSU and the National Design Award winner for Digital Design. She is a Co-Founder of Rael San Fratello and Emerging Objects, and her work merges traditional materials with advanced technologies. Her projects are widely published, and held in major museum collections, including MoMA, SFMOMA, and the Smithsonian.

Javier G. Fernandez is Professor of the Institute of Bioengineering of Catalonia at the University of Barcelona. His research is on the study of biological materials. At SUTD, he designed FLAM, a cellulosic bio-composite for large-scale manufacturing.

Stylianos Dritsas is Associate Professor in Architecture and Sustainable Design at the Singapore University of Technology and Design. His research is in design computation and digital fabrication. At SUTD, he developed the additive manufacturing process for FLAM.

Laia Mogas-Soldevila is the Director at DumoLab Research (DLR) and Graduate Architecture faculty at the Stuart Weitzman School of Design, University of Pennsylvania. She holds an interdisciplinary PhD in biomedical engineering and pioneers the field of biomaterial architectures by integrating life science practices into bio-fabricated material systems for design. Her laboratory develops water-based design and fabrication methods for responsive and regenerative building systems and everyday objects. Her work, awarded by the Johnson&Johnson Foundation, is recognized in global exhibitions as well as scientific journals, and influencing academic leadership across sustainable manufacturing, bio-based architecture, and biogenic materials innovation.

Paulina Grabowska is an architect and researcher specializing in circular design, biomaterials, biocomposites, computational design, and 3D printing. She previously worked at the BioMat department of the ITKE Institute at the University of Stuttgart, where she focused on developing chitin- and chitosan-based building materials and elements.

Hanaa Dahy, International Architect, Associate Professor in Sustainable Design at Aalborg University, Copenhagen, and former Professor at the University of Stuttgart, founded BioMat@ Copenhagen and BioMat GmbH and TGU@TTI GmbH (2022). A pioneer of "materials as a design tool," she led the architectural design of award-winning Smart Circular Bridge (DGNB 2025). Holder of European and international patents, she has received numerous design awards and delivered international lectures, including at Yale (2022), UVA (2023), and TUM (2023).

Christoph Zechmeister is an architect, computational designer, and researcher specializing in large-scale fiber composite structures. He earned degrees from TU Vienna and ETH Zürich, and completed his PhD at ICD, University of Stuttgart, where he leads the Computational Fiber Architecture research group within the Cluster of Excellence IntCDC.

Monika Göbel is Registered Architect, Senior Researcher, and Project Coordinator at ICD and IntCDC, University of Stuttgart. Her academic work focuses on integrative design, material systems, and fabrication. She has led key demonstrator projects, including the BUGA Pavilions, BUGA Wood Reuse, livMatS Biomimetic Shell, and the LAGA Wangen Projects.

Jan Knippers has led the Institute for Building Structures and Structural Design (ITKE) at the University of Stuttgart since 2000. His work focuses on innovative, resource-efficient structures, spanning research and practice. Key topics include advanced wood and fiber structures and bio-inspired compliant mechanisms.

Achim Menges is Registered Architect and Professor at the University of Stuttgart, founding director of ICD and IntCDC, and a Max Planck Fellow. He has held visiting professorships at Harvard GSD and internationally. In 2023, he received the Gottfried Wilhelm Leibniz Prize, and his design research is published and exhibited worldwide.

Mette Ramsgaard Thomsen is Professor of Digital Technologies and leads the Centre for Information Technology and Architecture (CITA), which she founded in 2005, at the Royal Danish Academy of Fine Arts, Schools of Architecture, Design and Conservation. Her research examines how computation is changing the material cultures of architecture.

Gabriella Rossi is Assistant Professor at CITA, Royal Danish Academy. Her research and teaching focuses on machine learning as an emerging modelling paradigm for AEC with focus on architectural datasets and their applications for sustainable materials, performative design, and adaptive robotic fabrication systems. She holds a PhD from the Royal Danish Academy.

Carl Eppinger is a PhD Candidate at CITA, Royal Danish Academy. His research focuses adaptive robotic fabrication for biopolymer composites, with the goal to facilitate their introduction to the construction industry, and joining adaptive hardware design with machine learning-based decision-making, creating a holistic fabrication approach.

Paul Nicholas is Associate Professor at CITA, at the Royal Danish Academy, and Director of the international masters program Computation in Architecture. Paul's research explores how advanced methods for predictive modeling and robotic fabrication with biomaterials can extend sustainable architectural practice and reframe how we build for the future.

ACKNOWLEDGMENTS

I would like to thank all the contributors in this volume, who generously shared their theoretical and experimental research within the theme of Plastic Futures in Architecture, and who granted permission to include their artworks in this project.

I also wish to express my sincere gratitude to the University of Virginia School of Architecture and the Jefferson Trust for their invaluable support in curating and organizing the Being Plastic/ Becoming Plastic Symposium and Exhibition held on March 2024, as well as in preparing this volume.

Special thanks to my student research assistants: Ipsita Datta, Avery Edson, Juliana Jackson, Eli Sobel, Tabi Summers, and Marisa Yamamoto, who assisted me throughout the research preparation and execution of the exhibition and this book.

Finally, I would like to thank my colleagues, mentors, and especially my friends and family for their continuous support.

1

PLASTIC TECHNE

An Introduction

Ehsan Baharlou

An Introduction

Not long ago, plastic, this *miraculous material*[1] with its mesmerizing beauty, was envisioned as the future of humanity. The promising future of a world with plastic now has engulfed us, and we are unable to imagine living without it.[2] A substance, with an essence of formability, turns into a suitable option for mass production. Modern plastics, as a synthetic polymer, replaced other conventional materials in production such as wood, metal, and ceramics.[3]

Within decades, synthetic polymers including various classes such as thermoplastics, thermosets, and elastomers have been incorporated into various industries, from cars and shipbuilding to household goods and building industries. The ease of its production allows industries to adjust their manufacturing lines to utilize plastic. Plastic manufacturing benefits from everlasting transformation, including casting, molding, pultrusion, extrusion, laminating, and winding.

In architecture, various synthetic polymers allow designers to transform their imaginative structures into the curvilinear form of the Monsanto House of the Future (1957) or create enchanting interior spaces like Verner Panton's Visiona II (1970). The new emergence of space inherits qualities of plasticity, without any limitations on color or form. This plastic space proclaims a new culture that transforms human identity into an obsession of using plastic products in daily life. Overconsumption of synthetic polymers poses a challenge to the natural ecosystem in which plastic waste becomes the defining feature of the Anthropocene, transforming the earth towards the Plasticene era. Humanity, as the sole inventor and consumer of this synthetic material, has an obligation to the future Plasticene.

The *Plastic Futures in Architecture* emphasizes on *Plastic Techne* to present ways of living with plastics in the Plasticene era. *Plastic Techne* refers to a state of construction in which plastic technology follows an obligation to comprehend the essence, process, and future of plastic materials in building. I refer to plastic as a noun and an adjective. Plastic, as a noun, refers to both synthetic- and natural-based polymers. As an adjective, it refers to the characteristics of generating fluid spaces with ultra-lightness, continuity, and malleability. Techne is not limited to the skill of production, in which the agency of a material moves beyond the entire process of

DOI: 10.4324/9781003613589-1

materialization. Techne is a thoughtful approach to technology where the material's end-of-life is part of the dematerialization.

After 70 years of modern plastic invention, such enchanting beauty is tainted with undeniable ugliness caused by its essence. This is its indelible characteristic that can rekindle its initial purpose with long-lasting life. Due to the long life of synthetic plastic, the balance between production and reclamation is disrupted, resulting in plastic pollution from macro to micro dimensions spreading in terraform, water, air, and even the human body.[4] We always believe recycling waste plastic can prevent us from becoming plastic. Plastics such as thermoplastics—categorized into seven types—are mostly recyclable, but elastomers are not when cured at increasing temperature. Currently, the recycled percentage of plastic waste is not promising, so we must change our attitude towards plastic use.

After the packaging industry, the building and construction industries produce the second-largest amount of plastic waste.[5] In other data analysis, the construction and building sector is responsible for approximately 40% of global carbon emissions, mostly through the extraction, processing, and production of building materials.[6,7] These challenges require the use of alternative building materials, or novel approaches such as Cradle-to-Cradle[8] and Circular Economy[9] methods, to reduce, reuse, or repurpose plastic waste residues. Understanding these challenges forces us to use a different mindset to explore ways of reclaiming plastic waste in building and construction sectors. The building life cycle increases the amount of time a building could be used as storage for embodied plastic waste. The process of downcycling plastic waste has been applied in the constructed environment; however, the toxicity and flammability of these substances must be adjusted to the specific building codes.

The second digital turn gradually established itself as a design means in the building sector, with approaches to prototyping any investigation for mitigating this polluting industry. When used to produce various prototypes, from scaled to one-to-one, synthetic plastic is examined as a building element. Proto-architecture movements are transforming the established applications of plastics to locate their reclamation potential in architecture. The technique of using digital production, integrated with computational design, extracts plastic morphology from synthetic polymers by which it can be interpreted as tracing forms and spaces created at the beginning of modern plastic invention.

Manufacturing techniques of plastic have been adopted to digital fabrication. From rapid prototyping of an architectural model to robotically winding building components, it shows how technological advancements help continue plasticity in architecture. According to the polluting potential of plastic products, the proto-architectural installations responsibly reinterpret the classic plastic beauty embedded in the works of artists, engineers, and architects, from Naum Gabo's work (Linear Construction in Space No.2, 1959–60) to a robotically wound ICD/ITKE research pavilion (livMatS Pavilion 2020–21). This responsible techne utilizes ecological materials, natural- or bio-based materials, or plastic reclamations to produce eco-friendly alternatives for shaping the constructed environment.

Plastic Techne moves beyond synthetic plastic reclamation to explore the natural polymers in design. Identifying the end-of-life of polymers and their natural degradation leads to a design process that provides building materials with a natural expiration period. In contrast, the industrialization of plastics was accompanied by valuing petroleum-based polymers over cellulose or other natural-based polymers.[10] We, designers and architects with a traditional understanding of natural polymers, are attempting to experiment with these materials again in the building industry.

Exploration of natural-based polymers emerges in an epoch where architectural designers experiment with natural polymers that have eco-friendly qualities. In academia, the collaboration between architects and other experts such as chemists, biologists, and engineers creates a deep understanding of natural-based polymers in the construction of proto-architectural artifacts. The importance of using natural-based polymers in architecture is the main source of these experiments. We are in the early stage of transcending synthetic plastic towards natural-based polymers. It might take time to convince other construction-related sectors to invest in the research and development of natural-based polymers for building materials.

In this volume, essays and case studies explore ways of living in the Plasticene era, highlighting examples of emergent plastic space, the development of materials and construction processes, and a transcendence to a new approach to plastic use. This volume builds upon the symposium and exhibition that was held at the University of Virginia School of Architecture in March 2024.

Plastic Futures in Architecture is organized in three main modes: Being, Becoming, and Beyond. *Being* raises awareness of the existence of plastic, how our identities have been shaped around built plastics, and how this material infiltrates our existing nature. *Becoming* refers to plastic technology advancement and how processing plastics has transformed our world. *Beyond* is a transcending state of synthetic polymers toward natural-based alternatives, which will result in a new epoch.

Notes

1 Roland Barthes, *Mythologies*, trans. Annette Lavers (Hill and Wang, 1972).
2 Jeffrey L. Meikle, *American Plastic: A Cultural History* (Rutgers University Press, 1995).
3 Susan Freinkel, *Plastic: A Toxic Love Story* (Houghton Mifflin Harcourt, 2011).
4 Roland Geyer et al., "Production, Use, and Fate of All Plastics Ever Made," *Science advances* 3, no. 7 (2017), https://doi.org/10.1126/sciadv.1700782.
5 OECD, "Global Plastics Outlook: Plastic Waste in 2019," OECD Environment Statistics (database), accessed February 25, 2026, https://doi.org/10.1787/a92f5ea3-en.
6 United Nations Environment Programme, "2020 Global Status Report for Buildings and Construction: Towards a Zero-emission, Efficient and Resilient Buildings and Construction Sector" (Nairobi, 2020).
7 World Green Building Council, "Advancing Net Zero Status Report 2020 World Green Building Council" (Toronto, 2020).
8 William McDonough and Michael Braungart, *Cradle to Cradle: Remaking the Way We Make Things*, 1st ed. (North Point Press, 2002).
9 Ellen MacArthur Foundation, "Towards the Circular Economy: Economic and business rationale for an accelerated transition" (2013).
10 Billie Faircloth, *Plastics Now: On Architecture's Relationship to a Continuously Emerging Material* (Routledge Taylor & Francis Group, 2015).

References

Barthes, Roland. *Mythologies*. Translated by Annette Lavers. Hill and Wang, 1972.

Ellen MacArthur Foundation. "Towards the Circular Economy: Economic and Business Rationale for an Accelerated Transition." 2013.

Faircloth, Billie. *Plastics Now: On Architecture's Relationship to a Continuously Emerging Material*. Routledge Taylor & Francis Group, 2015.

Freinkel, Susan. *Plastic: A Toxic Love Story*. Houghton Mifflin Harcourt, 2011.

Geyer, Roland, Jenna R. Jambeck, and Kara Lavender Law. "Production, Use, and Fate of All Plastics Ever Made." *Science Advances* 3, no. 7 (2017): e1700782. https://doi.org/10.1126/sciadv.1700782.

McDonough, William, and Michael Braungart. *Cradle to Cradle: Remaking the Way We Make Things*. 1st ed. North Point Press, 2002.

Meikle, Jeffrey L. *American Plastic: A Cultural History*. Rutgers University Press, 1995.

OECD. "Global Plastics Outlook: Plastic Waste in 2019." 2024. Accessed February 25, 2026. https://doi.org/10.1787/a92f5ea3-en.

United Nations Environment Programme. "2020 Global Status Report for Buildings and Construction: Towards a Zero-Emission, Efficient and Resilient Buildings and Construction Sector." Nairobi, 2020.

World Green Building Council. "Advancing Net Zero Status Report 2020 World Green Building Council." Toronto, 2020.

PART I
Being

Ehsan Baharlou

Being raises awareness of the pervasive presence of plastic in our environment. It explores how this human-made innovation changes our nature of existence and being and how synthetic polymers have been evolving and their central role in architecture and consequently in construction environment. This section highlights the challenges posed by plastic pollution and presets the innovative approaches to address these ecological challenges.

In *Being Plastic: Living in the Plasticene*, Phoebe Crisman notes that we live in Plasticene, where the planet earth is polluted with microplastics from air to water to soil, which changes the being of humanity, where the nano- or microplastics enter human bodies through inhalation and absorption. The chapter highlights the vital role of construction industry in the transformation toward Plasticene and how designers can explore creative approaches to mitigate this harmful transformation of the environment.

In *Plastic Imitating Life: Future Potential for a Controversial Material,* Blaine Brownell argues the life-like features of plastics from aesthetic capability to the self-healing and shape-shifting potential of plastics in construction and technology. The chapter points out how human-made polymers often overlook the ecological balance causing pollution and toxic waste, and highlights the future potential of improving recycling plastic waste and life-cycle focused approaches, such as biopolymers, which can offer a second chance to this controversial material.

A second chance might look like a heterogeneous system of coexisting and coevolving relationships among humans, nature, and technology, which is neither a utopian nor a dystopian space that can be created by the use of plastic waste to address the anthropogenic mass. In *Syntopia 0—Anthropos I Human: From "Designer Matter" to "Critical Material"*, Karola Dierichs presents an architectural project that critically responds to surpassing all living biomass by anthropogenic mass around 2020. The project examines critical materials theory through material design and construction robotics to create a spatial enclosure to reflect a monument of anthropogenic mass from designed granular materials made of injection molding of plastic waste recycler form car industries.

In *Breathing Enclosures and Engineered Living Materials*, Maria Paz Gutierrez discusses the emergence of Engineered Living Materials (ELMs) for the future of construction. The chapter explores the integration of biosynthesis and materials science within the context of ELMs to

DOI: 10.4324/9781003613589-2

create building enclosure with metabolic functions in contrast to the mechanistic model of synthetic polymers. A project featuring lichen-laden hydrogel biowall system exemplifies the integration of metabolic function and multifunctional biopolymers in developing a membrane for detoxifying air and improving indoor air quality.

2

BEING PLASTIC

Living in the Plasticene

Phoebe Crisman

Today we live in the Plasticene amidst plastic soil, plastic water, plastic air, and plastic food. Microplastics are found from the top of Mount Everest[1] to the deepest ocean trench.[2] Synthetic plastic is unequivocally one of the most widely produced, used, and wasted materials of modern times. Over 99% of that plastic is made from fossil fuel-derived chemicals.[3] While single-use plastic is perhaps the biggest challenge, contemporary buildings and other durable goods are comprised of vast amounts of plastics as well. Our relatively recent addiction to synthetic plastic, since the 1970s, has produced severe environmental, socio-economic, and health consequences. We all know that the proliferation of fossil fuel-based plastics on the planet is a truly a wicked global problem—but how did we get here? And where should we go?

Plastic before 1950

The first fully synthetic plastic polymer, meaning it contained no molecules found in nature, was invented and received a US patent in 1907 by Leo Baekeland. However, humans have used many natural polymers such as rubber, lacquer, shellac, and horn drawn from plant, insect, and animal sources since antiquity. In the mid-nineteenth century, other natural polymers such as linoleum were created by combining linseed oil, sawdust, and other naturally occurring materials. To meet the growing demand for materials that could be plastically molded into durable forms, chemists then produced a wide range of semi-synthetic plastics by chemically modifying natural polymers. Parkesine, the first semi-synthetic plastic, was displayed at the London International Exhibition of Industry and Art in 1862. The exhibition guide emphasized the novelty of this material: "Amongst the most extraordinary substances shown is a new material called Parkesine ... a substance hard as horn, but as flexible as leather, capable of being cast or stamped, painted, dyed or carved."[4] In the 1930s, German chemist Hermann Staudinger and American chemist Wallace Carothers made major advancements in polymer science, resulting in the development of several thermoplastics, including polyethylene, polyurethane, and polystyrene.[5]

Architects and industrial designers promptly embraced the new formal and spatial possibilities of these seemingly magical substances. In "Plastics: The Building Blocks of the Twentieth Century," historian Anthony Walker examined the emergence of these new plastic products and their image

DOI: 10.4324/9781003613589-3

of plasticity, lightness, and transparency. Architects began to use laminated sheet plastics for countertops, doors, bathroom wall surfaces, and other interior applications in the 1930s. There were radical experiments, like the Vinylite 'house'—three rooms constructed entirely of vinyl—at Chicago's Century of Progress International Exposition in 1933.[6] Architectural innovations in plastic moved from simple surfaces to lightweight glass reinforced plastic (GRP, also known as fiberglass) structural panels, complex shells, and folded plate forms.

The Mid-Century Magic of Plastic

To understand why and how plastic proliferated in so many sectors, it is crucial to acknowledge how mid-twentieth century consumer culture embraced the newness, beauty, and magical qualities of plastic. Roland Barthes' seminal 1957 essay "Plastic" is a good place to start.

> Despite having names of Greek shepherds (Polystyrene, Polyvinyl, Polyethylene), plastic ... is in essence the stuff of alchemy. ... So, more than a substance, plastic is the very idea of its infinite transformation; as its everyday name indicates, it is ubiquity made visible. And it is this, in fact, which makes it a miraculous substance: a miracle is always a sudden transformation of nature. Plastic remains impregnated throughout with this wonder: it is less a thing than the trace of a movement.[7]

Barthes observed the uncanny transmutation of matter, and particularly the movement embodied in the shaping of the plastic thing. But now we understand the terrifying magic of infinite subdivision and global movement of invisible plastic particles. Reflecting on Barthes' reference to movement rather than objecthood, Meera Subramanian described how "Plastics are never just objects; they're bodies migrating within ecosystems and through other, larger bodies."[8] Today microplastics (MPs) migrate through all living creatures. As Barthes anticipated nearly seventy years ago, "The hierarchy of substances is abolished: a single one replaces them all: the whole world *can* be plasticized, and even life itself since, we are told, they are beginning to make plastic aortas."[9]

In *Plastic Matter,* Heather Davis interrogates plastic from a socio-economic viewpoint. She argues that "plastic reveals broader assumptions about relations to matter, and how matter is understood under technocapitalism. Plastic matter describes the assumptions that matter is there to be manipulated; it can and should be bent and made pliable; and its potential for manipulation is endless."[10] She links this pliability to the post-Enlightenment project. The development of various types of plastic has allowed us to seemingly control nature and specific natural properties in unprecedented ways. Like Barthes—who asserted that the "scope' of [plastic's] transformations gives man the measure of power, and since the very itinerary of plastic gives him the euphoria of a prestigious free-wheeling through Nature"—Davis emphasizes that "[p]lastics are *engineered* to feel universal. They are possessed of, or they are designed to want, the quality of universality."[11] Indeed, plastic is "the perfect material of the modern age's dream of human mastery over, and separation from, the earth: infinitely reproducible, down to the molecular level."[12] Davis observes that "Plastics were the beginning of a long line of inventions which rest on the premise that ideal matter can be created through scientific and corporate technological practices."[13] The rise of advertising contributed to the idea of control over consumers as well.

Advertising and trade organizations played an essential role in the promotion and rapid adoption of plastic products. For instance, Barthes wrote his seminal essay after attending a Plastics trade show and observing the seemingly magical transformation of small plastic pellets into final objects. "At one end, raw, telluric matter, at the other, the finished, human object; and between these two extremes, nothing; nothing but a transit, hardly watched over by an attendant in a cloth cap, half-god, half-robot."[14] Advertisers exploited this incomprehensible technological transmutation in their messages to consumers. A 1950s advertisement for Dow's Saran Wrap product, literally refers to magic. "Look! It's crystal clear and it clings like magic!"[15] A 1953 Dupont cellophane ad claims that "The best things in life come in cellophane" and depicted a human baby wrapped in plastic.[16] Along with single use packaging, consumers were encouraged to replace glass and ceramic dishes and other durable items with plastic substitutes.[17] Throw away culture and convenience was the central message.

In *Vibrant Matter: A Political Ecology of Things,* Jane Bennett examines the 'thing power' of objects as "vivid entities not entirely reducible to the contexts in which (human) subjects set them" and how even debris has a live presence.[18] She notes that "American materialism, which requires buying ever-increasing numbers of products purchased in ever-shorter cycles, is antimateriality. The sheer volume of commodities, and the hyperconsumptive necessity of junking them to make room for new ones, conceals the vitality of matter."[19] In the case of plastic, its ubiquity devalues its materiality and the possibility of functional longevity. And of course, those advertisements did not convey the long lifespan of plastics and their serious negative impacts—from fossil fuel extraction to worker health to solid waste and MPs. Spurred by Rachel Carson's 1962 book *Silent Spring* and a growing questioning of chemicals that had entered our daily lives, public concerns about the performance limitations and toxicity of plastic emerged by the late 1960s and early 1970s.[20]

Many types of synthetic plastic appeared in both typical building construction and high-profile architecture in the mid-twentieth century. GRP and transparent plexiglass (PMMA) technology continued to develop and inspire architects designing World Fair pavilions in partnership with plastics manufacturers. Described by the *Architectural Forum* as "Ed Stone's plastic palace," architect Edward Durrell Stone's US Pavilion at the 1958 Brussels World's Fair used *Kalwall,* a then experimental system of translucent plastic and aluminum, to clad a 340-foot diameter free-span, cable-tensioned roof structure.[21] The lightweight plastic cladding filled the massive circular space with sunlight and allowed an incredibly minimal, bicycle wheel-like structure. GRP was widely used at the 1964 New York World's Fair as well. Building on Stone's work, architect Phillip Johnson's massive (350' × 250') elliptical New York State pavilion—the Tent of Tomorrow—employed colorful plastic Kalwall sheeting.[22] Architect Peter Schladermundt's three World's Fair Marina pavilions embraced fluidity of fiberglass to create dynamic shell structures.[23] Plastic cladding quickly moved out of the Fairs into commercial buildings. For example, British architect James Stirling explored the image of plasticity in the curvaceous exterior GRP cladding of the Olivetti Training Centre in 1969.[24] During the 1960s and 1970s, architects explored new forms of space often using polyvinyl chloride (PVC) to create inflatable architecture. However, because many plastics could not achieve adequate fire ratings, their use was more limited in building construction than industrial design.

Post-1970s Plastic Production and Waste

Since the 1970s, the rate of plastic production has grown faster than any other material. Figure 2.1 from the Organisation for Economic Co-Operation and Development 2024 report, *Policy Scenarios of Eliminating Plastic Pollution by 2040,* depicts this dramatic growth. Compared to current levels, only a small amount of plastic was produced in the 1950s—1.5 million tons

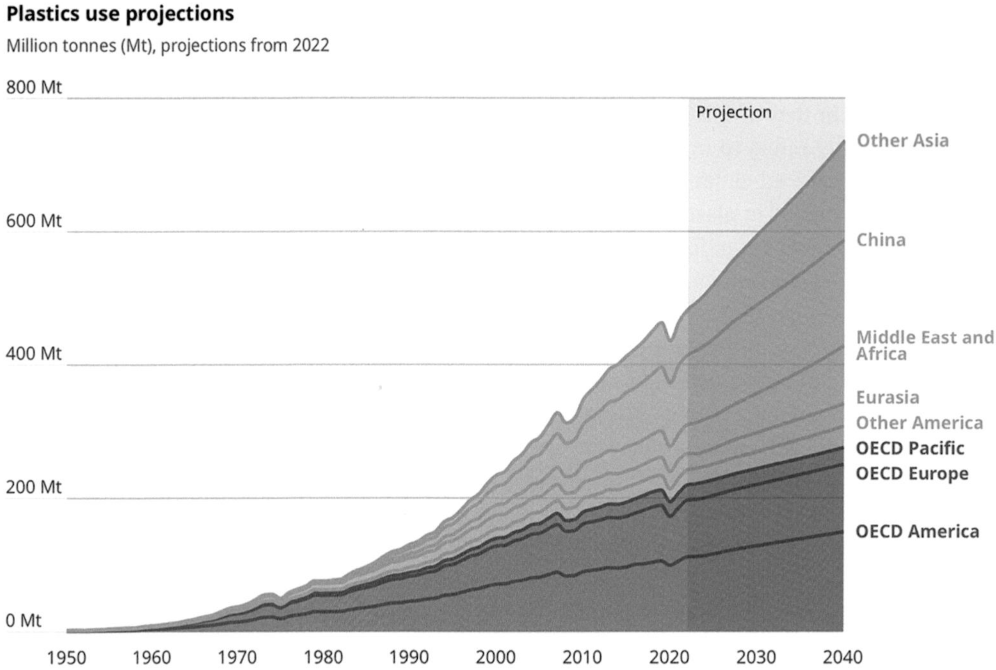

FIGURE 2.1 Plastic use projections.

Source: Policy Scenarios of Eliminating Plastic Pollution by 2040, OECD, 2024.

annually.[25] However, between 2000 and 2019, the global production of plastic rose from 234 to 460 million tons (Mt).[26] How can we even comprehend the physical reality of this number? If these trends continue, “Annual plastics production and use is projected to rise from 435 million tonnes (Mt) in 2020 to 736 Mt in 2040.”[27]

In addition to the sheer quantity of production, the longevity of plastic as waste is concerning. In the early 2000s, the amount of plastic waste generated rose more in ten years than in the previous forty. Packaging produces the greatest amount of plastic waste, followed by clothing and consumer products. The longevity of plastic, once a primary benefit, was now seen as a threat to ecosystems and our own human bodies. Plastic recycling was identified as a solution to plastic waste in the early 1970s, but in 2024 the OECD reported that only 6% of plastic is successfully recycled globally.[28] Half of all synthetic plastic produced is designed to be used just once and then thrown away as single-use plastic. In places where solid waste management is inadequate or non-existent, this plastic material flows into waterways, forms ocean gyres, and washes onto beaches as marine debris. The United Nations now measures the annual “emission” of rivers into the ocean, and how plastic is transported by currents globally.

The Plasticene

Plastics are now part of the Earth’s fossil record and a marker of the Anthropocene—our current geological epoch in which human activities have come to dominate the planet. Exactly where and when did the Anthropocene begin? Scientists have identified many markers: radioactive isotopes produced by cold war H-bomb tests, tiny spheroidal carbonaceous particles[29] only produced by the

FIGURE 2.2 Plastic beach sand on Syros island, Greece, 2025. Photograph by Michael Petrus.

high-temperature burning of coal or heavy oil, chemical and metal pollution, the huge amount of broiler chicken bones, and of course—plastic.[30] In 2019, scientists found that plastic was deposited into the earth's strata and argued that the stone age and iron age have been followed by the plastic age—the Plasticene. Plastiglomerate, a new stone "formed through intermingling of melted plastic, beach sediment, basaltic lava fragments, and organic debris," is now found on beaches around the world and discussed as a possible marker of the Anthropocene.[31] Figure 2.2 is a photograph from a remote Greek beach that I visited in 2024. The beach seemed free of plastic trash, but upon closer examination the sand itself was plastic. Non-human creatures unknowingly consume large quantities of plastics that lead to sickness and death. Microbiologist Linda Amaral-Zettler named a new marine microbial habitat the "plastisphere"—where microscopic life thrives on tiny floating islands of debris and can be transported great distances.[32] Plastic waste is fully integrated within biological and ecological systems. "As plastics degrade, so do the barriers that once seemed so defined and distinct, between inert and organic, between outside a body and its interior, between science and art, between present and future."[33]

Microplastics and Nanoplastics

MPs, synthetic plastic particles less than 5 mm across, and nanoplastics, smaller than 1 micrometer (μm) in size, are a hazard found in deep-sea sediments, rain, and the air we breathe.[34] Primary MPs are specifically produced to be tiny, such as nurdles—plastic pellets used to make plastic objects, and the microbeads still used in cosmetics in some countries. Secondary MPs are generated when larger plastic pieces break down due to mechanical strain, hydrolysis, oxidation and weathering effects from UV light, heat, and microorganisms.[35] According to The Handbook of Environmental Chemistry, "It is estimated that secondary sources of MPs currently account for the dominant of MPs in the environment although large plastic wastes need hundreds of years to break down into MPs under natural conditions."[36]

Plastic now falls from the sky in the rain; it's literally raining plastic. In a 2020 study published in *Science,* rainwater and air samples were collected in 11 remote protected conservation areas in the western United States. Researchers found that over 1,000 metric tons of microplastic particles fall onto that remote area each year. That's the equivalent of over 120 million plastic water bottles.[37] Their study area included only 6% of the total US land area. Imagine the amount of microplastic particles that fall on our suburbs, cities, and farmlands. Between 66% and 70% of those plastic particles were synthetic microfibers from sources like polyester clothing and carpeting. "Approximately 30% of the particles were primary plastic microbeads ranging in size from 5 to 30 μm in a wide variety of colors"—mostly from industrial paints and coatings that are a major concern for the building sector.[38] MPs are transported via air and water around the world—far more insidious than acid rain and unable to be removed from our environment.

We are all plastic now, as swirling micro- and nanoplastics enter our bodies through inhalation and absorption. MPs accumulate in our tissues and organs, including our lungs, liver, spleen, and kidneys.[39] They are found in our blood, mother's milk, and the placentas of newborn babies.[40] The bioaccumulation of MPs has even been found in the brains of deceased humans.[41] "A new study by the University of Newcastle, Australia suggests that an average person could be ingesting approximately 5 grams of plastic every week. The equivalent of a credit card's worth of microplastics."[42] Figure 2.3 depicts the percentage of tap water samples containing plastic fibers, and the average number of fibers per 500 ml. These numbers vary significantly by country and region. For example, 94% of water samples contained plastic fibers in the United States compared with 72.2% in Europe. There is substantial evidence that plastic-associated chemicals, such as methyl mercury, plasticizers, and flame retardants, enter our bodies and are linked to health concerns. While scientists don't know exactly what inhaling or ingesting MPs mean for human health, they know it cannot be good.[43]

Our perception of plastic has turned from fascination to fear. In "Polymers are Forever," Alan Weisman noted

> no plastic has died a natural death yet. It took today's microbes that break hydrocarbons down to their building blocks a long time after plants appeared to learn to eat lignin and cellulose. More recently, they've even learned to eat oil. None can digest plastic yet, because fifty years is too short a time for evolution to develop the necessary biochemistry.[44]

Yet, recent science is advancing in this area.[45] Microbiologists have discovered a single species of bacteria that breaks down polyethylene terephthalate (PET), but at a very slow pace.[46] There is much research to be done before a large-scale microbial solution could be viable. As Mark O'Connell wrote in *Notes from an Apocalypse: A Personal Journey to the End of the World and Back:*

> The whole subject of microplastics is possessed of a nightmarish lucidity, because we understand it to be a symptom of a deeper disease. The unthinkable harm we have done to

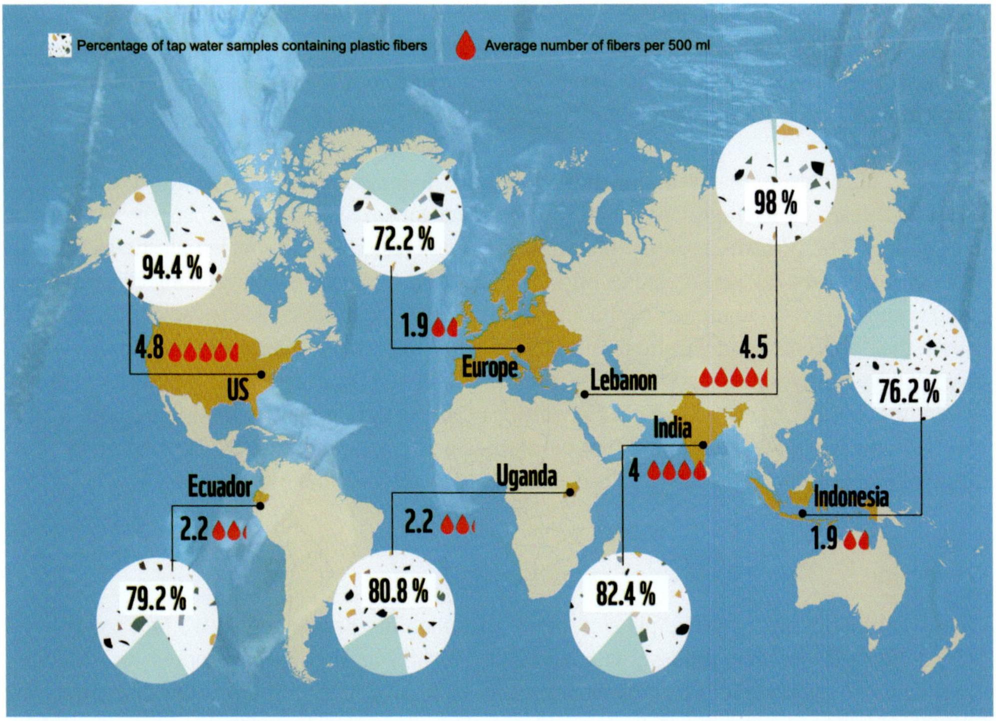

FIGURE 2.3 Map of average percentage of tap water samples containing plastic fibers and average number of fibers.

Source: No Plastic in Nature: Assessing Plastic Ingestions from Nature to People. World Wildlife Fund, 2019.

the planet—that is done to the planet on our behalf, as consumers—is being visited, in this surreal and lurid manner, on our own bodies.[47]

Synthetic Plastics in Contemporary Architecture

So how does this all relate to architecture? The construction industry, via construction and demolition waste (CDW), "accounts for more than a third of all waste generated in the EU."[48] This waste stream contains a significant quantity of plastics, along with metals, concrete, bricks, wood, and glass. Construction consumes 14% of yearly plastic output.[49] Sometimes this number is listed as 20%, but either way, that's a lot of plastic in our buildings and MPs in our environment. Plastics are the default choice used in a wide range of building components, such as roofing, membranes, formwork, building skins, windows and doors, seals, gutters, insulation, pipe and cables, and internal and external finishes, including flooring, paints, varnishes, and stains. In most cities and town across the United States, some components, such as plumbing pipe and electrical conduit, are mandated by building codes to be plastic. However, there are many cases where we can consider returning to materials that had been used pre-plastic. For instance, choosing metal gutters, typically made from recycled steel, is a more durable choice than plastic. While the initial cost is slightly more, the return on investment is greater. Fabricating window frames from wood rather than vinyl is ultimately a more durable, maintainable, and beautiful solution. The harmful effects of the production of vinyl, on both worker and nearby communities, is a topic unto itself. Thinking in systems and life cycle assessment requires that we think about the externalities in the production of each of these plastic components, from fossil fuel and other raw material extraction, through production, use, and material end of life in a landfill.

The good news is that building materials have a longer lifespan than most plastics ... thirty-five years on average, compared to five years for textiles and less than six months for packaging.[50] Many plastic building materials are now created from recycled plastic as well. In some cases, however, well-meaning efforts to create a circular economy using plastic waste streams can pose risks to human health and well-being in the process. Recent scientific research shows that products made from plastic waste can concentrate chemicals and toxins—thus potentially amplifying adverse health impacts.[51] Some scholars argue that a short-term focus on repurposing large amounts of plastic waste can inadvertently contribute to long-term health crises. Other scientists have found that bioplastics and plant-based materials often make little difference to toxicity as the same additives are used to achieve material performance.[52]

Conclusion

Like fossil fuel-based synthetic plastics, many sustainability problems that we face today have been deliberately or thoughtlessly created since World War II. Exurban sprawl, automobile-dependent life, globalized manufacturing, and excessive consumption—the list goes on. Plastics addiction is another self-made problem based on a technology developed for the instant gratification of convenience and massive profit. We made this tragedy in 75 years—but there are alternatives. This book contains many encouraging new ideas for bioplastic alternatives to synthetic plastic. We can change course if we pursue a path that requires more creative and holistic thinking. It is perhaps less convenient in the short term, but we need to adopt a more thoughtful path that values abundant life over the slow asphyxiation of all living things by plastics.

Notes

1 Imogen Napper et al., "Reaching New Heights in Plastic Pollution—Preliminary Findings of Microplastics on Mount Everest," *One Earth* 3, issue 5 (2020): 621–630. https://doi.org/10.1016/j.oneear.2020.10.020.
2 X. Peng et al., "Microplastics Contaminate the Deepest Part of the World's Ocean," *Geochemical Perspectives Letters* 9 (2018): 1–5. https://doi.org/10.7185/geochemlet.1829.
3 "Fossil Fuels & Plastic," Center for International Environmental Law, accessed August 15, 2025, www.ciel.org/issue/fossil-fuels-plastic/.
4 Morris Kaufman, *The First Century of Plastics: Celluloid and Its Sequel.* (Plastics Institute, 1963), 16.
5 Anthony Walker, "Plastics: The Building Blocks of the Twentieth Century," *Construction History* 10 (1994): 67. www.jstor.org/stable/41613731.
6 Walker, "Plastics," 72.
7 Roland Barthes, "Plastic," in *Mythologies,* trans. Annette Lavers (Hill and Wang, 1972): 97. This article was originally written between 1954 and 1956 and reprinted as "Le Plastique" in *Mythologies* (Editions du Seuil, 1957): 192–94.
8 Meera Subramanian, "The Nature of Plastics: Explorations at the edge of the artificial," *Orion* (Spring 2021).
9 Barthes, *Mythologies*, 99.
10 Heather Davis, *Plastic Matter* (Duke University Press, 2022), 1.

11 Davis, *Plastic Matter,* 47.
12 Davis, *Plastic Matter,* 48.
13 Davis, *Plastic Matter,* 47.
14 Barthes, *Mythologies*, 97.
15 "1953 Saran Wrap ad," Dow, accessed August 15, 2025, https://corporate.dow.com/en-us/about-dow/company/history/vintage-chemical-advertising.html.
16 Cellophane, Du Pont, "The best things in life come in Cellophane," advertisement, *Saturday Evening Post,* August 29, 1953, 101.
17 Tom Fisher, "A World of Colour and Bright Shining Surfaces: Experiences of Plastics after the Second World War," *Journal of Design History* 26, no. 3 (2013): 285–303. www.jstor.org/stable/43831508.
18 Jane Bennett, *Vibrant Matter: A Political Ecology of Things* (Duke University Press, 2010), 5.
19 Bennett, *Vibrant Matter,* 5.
20 Susan Freinkel, *Plastic: A Toxic Love Story* (Houghton Mifflin Harcourt, 2011).
21 Edward Durell Stone, "An Architecture of Space and Grace," *Architectural Record* 122, no. 1 (July 1957): 156.
22 Mildred F. Schmertz, "Architecture at the New York World's Fair," *Architectural Record* 136 (July 1964): 144.
23 Mary Jablonsky, "Fiberglass Reinforced Plastic at the Fair," in *Mid-Century Modern Structures: Materials and Preservation Symposium Proceedings* (National Park Service, April 2015). Also, Kirsten Hively, "The Candela Structures: Architecture as Storytelling," *Urban Omnibus,* (June 30, 2010), accessed June 20, 2025. https://urbanomnibus.net/2010/06/the-candela-structures-architecture-as-storytelling/.
24 Walker, "Plastics," 75–80.
25 Plastics Europe Market Research Group, *The Compelling Facts about Plastics* (Plastics Europe, January 2008).
26 OECD, *Global Plastics Outlook: Economic Drivers, Environmental Impacts and Policy Options* (OECD Publishing, 2022): 14. https://doi.org/10.1787/de747aef-en.
27 OECD, *Policy Scenarios of Eliminating Plastic Pollution by 2040* (OECD Publishing, 2024): 11. https://doi.org/10.1787/76400890-en.
28 OECD, "*Policy Scenarios,"* 11.
29 Graeme T. Swindles et al., "Spheroidal carbonaceous particles are a defining stratigraphic marker for the Anthropocene," *Scientific Reports* 5, 10264 (2015). https://doi.org/10.1038/srep10264.
30 Damien Carrington, "After bronze and iron, welcome to the plastic age, say scientists," *The Guardian,* September 4, 2019.
31 Patricia L. Corcoran et al., "An anthropogenic marker horizon in the future rock record," *GSA Today* 24, issue 6 (June 2014): 4–8. https://doi.org/10.1130/GSAT-G198A.1.
32 Linda A. Amaral-Zettler et al., "The biogeography of the Plastisphere: implications for policy," *Frontiers in Ecology and the Environment* 13, issue 10 (December 2015): 541–546. https://doi.org/10.1890/150017.
33 Subramanian, "The Nature of Plastics," 66.
34 Lisbeth Van Cauwenberghe et al., "Microplastic pollution in deep-sea sediments," *Environmental Pollution* 182 (2013): 495–499. https://doi.org/10.1016/j.envpol.2013.08.013.
35 Lapyote Prasittisopin et al., "Microplastics in Construction and Built Environment," *Developments in the Built Environment* 15 (October 2023): 1. https://doi.org/10.1016/j.dibe.2023.100188.
36 L. An et al., "Sources of Microplastic in the Environment. In *Microplastics in Terrestrial Environments. The Handbook of Environmental Chemistry,* eds He, D. and Luo, Y. (Springer, Cham, 2020). https://doi.org/10.1007/698_2020_449.
37 Matt Simon, "Plastic Rain is the New Acid Rain," *Wired* (June 11, 2020).
38 Janice Brahney et al., "Plastic rain in protected areas of the United States," *Science* 368, no. 6496 (June 12, 2020): 1257–1260. https://doi.org/10.1126/science.aaz5819.
39 Nur Sakina Roslan et al., "Detection of microplastics in human tissues and organs: A scoping review," *Journal of Global Health* 14.04179 (2004): 1-15. https://doi.org/10.7189/jogh.14.04179.
40 Heather Leslie et al., "Discovery and quantification of plastic particle pollution in human blood," *Environmental International* 163 (May 2022). Also, Damian Carrington, "Microplastics found in human blood for first time," *The Guardian* (March 24, 2022).
41 Alexander Nihart et al., "Bioaccumulation of microplastics in decedent human brains," *Nature Medicine* 31 (2025): 1114–1119. https://doi.org/10.1038/s41591-024-03453-1

42 Dalberg and The University of Newcastle, Australia, *No Plastic in Nature: Assessing Plastic Ingestions from Nature to People* (World Wildlife Fund, 2019): 6.
43 Yufei Wang and Haifeng Quian, "Phthalates and their Impacts on Human Health," *Healthcare* 9.603 (2021): 1–9. https://doi.org/10.3390/healthcare9050603.
44 Alan Weisman, "Polymers are Forever," *Orion* (May/June 2007).
45 Stephen Buranyi, "'We are just getting started': the plastic-eating bacteria that could change the world," *The Guardian* (September 28, 2023).
46 Shosuke, Yoshida et al. "A bacterium that degrades and assimilates poly(ethylene terephthalate)," *Science* 351, no. 6278 (March 11, 2016): 1196–1199. https://doi.org/10.1126/science.aad6359.
47 Mark O'Connell, *Notes from an Apocalypse: A Personal Journey to the End of the World and Back* (Doubleday, 2020).
48 "*Construction and Demolition Waste,"* European Commission, accessed July 20, 2025, https://environment.ec.europa.eu/topics/waste-and-recycling/construction-and-demolition-waste_en. Also see Norman Miller, "The Industry Creating a Third of the World's Waste," *BBC,* December 15, 2021. www.bbc.com/future/article/20211215-the-buildings-made-from-rubbish.
49 Prasittisopin et al., "Microplastics," 2.
50 Roland Geyer et al., "Production, use, and fate of all plastics ever made," *Science Advances* 3, no. 7 (July 19, 2017). https://doi.org/10.1126/sciadv.1700782.
51 O. Horodytska et al., "Non-intentionally added substances (NIAS) in recycled plastics," *Chemosphere* 251 (July 2020). https://doi.org/10.1016/j.chemosphere.2020.126373
52 Lisa Zimmermann et al., "Are bioplastics and plant-based materials safer than conventional plastics? *In vitro* toxicity and chemical composition," *Environment International* 145 (December 2020). https://doi.org/10.1016/j.envint.2020.106066.

References

Amaral-Zettler, Linda, Erik Zettler, Beth Slikas, et al. "The Biogeography of the Plastisphere: Implications for Policy." *Frontiers in Ecology and the Environment* 13, no. 10 (December 2015): 541–546. https://doi.org/10.1890/150017.

An, L., Q. Liu, Y. Deng, et al. "Sources of Microplastic in the Environment." In *Microplastics in Terrestrial Environments. The Handbook of Environmental Chemistry,* eds He, D. and Luo, Y. Springer, 2020. https://doi.org/10.1007/698_2020_449.

Stone, Edward Durell. "An Architecture of Space and Grace." *Architectural Record* 122, no. 1 (July 1957): 153–176.

Barthes, Roland. "Plastic." In *Mythologies,* trans. Annette Lavers. Hill and Wang, 1972.

Bennett, Jane. *Vibrant Matter: A Political Ecology of Things.* Duke University Press, 2010.

Brahney, Janice, Margaret Hallerud, Eric Heim, et al. "Plastic Rain in Protected Areas of the United States." *Science* 368, no. 6496 (June 12, 2020): 1257–1260. https://doi.org/10.1126/science.aaz5819.

Buranyi, Stephen. "'We Are Just Getting Started': The Plastic-Eating Bacteria that Could Change the World." *The Guardian*, September 28, 2023.

Carrington, Damien. "After Bronze and Iron, Welcome to the Plastic Age, Say Scientists." *The Guardian,* September 4, 2019.

Cellophane. Du Pont. "The Best Things in Life Come in Cellophane." *Advertisement. Saturday Evening Post,* August 29, 1953.

Center for International Environmental Law. "Fossil Fuels & Plastic." Accessed August 15, 2025. www.ciel.org/issue/fossil-fuels-plastic/.

Corcoran, Patricia L., Charles J. Moore, Kelly Jazvac. "An Anthropogenic Marker Horizon in the Future Rock Record." *GSA Today* 24, no. 6 (June 2014): 4–8. https://doi.org/10.1130/GSAT-G198A.1.

Dalberg and The University of Newcastle, Australia. *No Plastic in Nature: Assessing Plastic Ingestions from Nature to People.* World Wildlife Fund, 2019.

Davis, Heather. *Plastic Matter.* Duke University Press, 2022.

Dow. "1953 Saran Wrap ad." Accessed August 15, 2025. https://corporate.dow.com/en-us/about-dow/company/history/vintage-chemical-advertising.html.

European Commission. "Construction and Demolition Waste." Accessed July 20, 2025. https://environment.ec.europa.eu/topics/waste-and-recycling/construction-and-demolition-waste_en.

Fisher, Tom. "A World of Colour and Bright Shining Surfaces: Experiences of Plastics after the Second World War." *Journal of Design History* 26, no. 3 (2013): 285–303. www.jstor.org/stable/43831508.

Freinkel, Susan. *Plastic: A Toxic Love Story.* Houghton Mifflin Harcourt, 2011.

Geyer, Roland, Jenna Jambeck, Kara Lavender Law. "Production, Use, and Fate of All Plastics Ever Made." *Science Advances* 3, no. 7 (July 19, 2017). https://doi.org/10.1126/sciadv.1700782.

Hively, Kirsten. "The Candela Structures: Architecture as Storytelling." *Urban Omnibus,* (June 30, 2010). *Accessed* June 20, 2025. https://urbanomnibus.net/2010/06/the-candela-structures-architecture-as-storytelling/.

Horodytska, O., A. Cabanes, A. Fullana. "Non-Intentionally Added Substances (NIAS) in Recycled Plastics." *Chemosphere* 251 (July 2020). https://doi.org/10.1016/j.chemosphere.2020.126373.

Jablonsky, Mary. "Fiberglass Reinforced Plastic at the Fair." In *Mid-Century Modern Structures: Materials and Preservation Symposium Proceedings.* National Park Service, April 2015.

Kaufman, Morris. *The First Century of Plastics: Celluloid and Its Sequel.* Plastics Institute, 1963.

Leslie, Heather, Martin van Velzen, Sicco Brandsma, et.al. "Discovery and Quantification of Plastic Particle Pollution in Human Blood." *Environmental International* 163 (May 2022): 107199. https://doi.org/10.1016/j.envint.2022.107199.

Miller, Norman. "The Industry Creating a Third of the World's Waste." *BBC,* December 15, 2021. www.bbc.com/future/article/20211215-the-buildings-made-from-rubbish.

Napper, Imogen, Bede Davis, Heather Clifford, et al. "Reaching New Heights in Plastic Pollution—Preliminary Findings of Microplastics on Mount Everest." *One Earth* 3, no. 5 (2020): 621–630. https://doi.org/10.1016/j.oneear.2020.10.020.

Nihart, A. J., M. A. Garcia, E. El Hayek, et al. "Bioaccumulation of Microplastics in Decedent Human Brains." *Nature Medicine* 31 (2025): 1114–1119. https://doi.org/10.1038/s41591-024-03453-1.

O'Connell, Mark. *Notes from an Apocalypse: A Personal Journey to the End of the World and Back.* Doubleday, 2020.

OECD. *Global Plastics Outlook: Economic Drivers, Environmental Impacts and Policy Options.* OECD Publishing, 2022. https://doi.org/10.1787/de747aef-en.

OECD. *Policy Scenarios of Eliminating Plastic Pollution by 2040.* OECD Publishing, 2024. https://doi.org/10.1787/76400890-en.

Peng, X., M. Chen, S. Chen, et.al. "Microplastics Contaminate the Deepest Part of the World's Ocean." *Geochemical Perspectives Letters* 9 (2018): 1–5. https://doi.org/10.7185/geochemlet.1829.

Plastics Europe Market Research Group. *The Compelling Facts about Plastics.* Plastics Europe, January 2008.

Prasittisopin, Lapyote, Wahid Ferdous, Viroon Kamchoom. "Microplastics in Construction and Built Environment." *Developments in the Built Environment* 15 (October 2023). https://doi.org/10.1016/j.dibe.2023.100188.

Roslan, Nur Sakina, Y. Y. Lee, Y. S. Ibrahim, et al. "Detection of Microplastics in Human Tissues and Organs: A Scoping Review." *Journal of Global Health* 14 (2004): 1–15, 04179. https://doi.org/10.7189/jogh.14.04179.

Schmertz, Mildred F. "Architecture at the New York World's Fair." *Architectural Record* 136 (July 1964): 143–150.

Simon, Matt. "Plastic Rain Is the New Acid Rain." *Wired,* June 11, 2020.

Subramanian, Meera. "The Nature of Plastics: Explorations at the Edge of the Artificial." *Orion,* (Spring 2021).

Swindles, G., E. Watson, T. Turner, et al. "Spheroidal Carbonaceous Particles Are a Defining Stratigraphic Marker for the Anthropocene." *Scientific Reports* 5 (2015): 10264. https://doi.org/10.1038/srep10264.

Van Cauwenberghe, Lisbeth, Ann Vanreusel, Jan Mees, et al. "Microplastic Pollution in Deep-Sea Sediments." *Environmental Pollution* 182 (2013): 495–499. https://doi.org/10.1016/j.envpol.2013.08.013.

Walker, Anthony. “Plastics: The Building Blocks of the Twentieth Century.” *Construction History* 10 (1994): 67–88. www.jstor.org/stable/41613731.

Wang, Yufei, Haifeng Quian. “Phthalates and Their Impacts on Human Health.” *Healthcare* 9 (2021): 1–9, 603. https://doi.org/10.3390/healthcare9050603.

Weisman, Alan. “Polymers are Forever.” *Orion,* (May/June 2007).

Yoshida, Shosuke, Kazumi Hiraga, Toshihiko Takehana et al. “A Bacterium that Degrades and Assimilates Poly(Ethylene Terephthalate).” *Science* 351, no. 6278 (March 11, 2016): 1196–1199. https://doi.org/10.1126/science.aad6359.

Zimmermann, Lisa, Andrea Dombrowski, Carolin Volker, et al. “Are Bioplastics and Plant-Based Materials Safer Than Conventional Plastics? *In vitro* Toxicity and Chemical Composition.” *Environment International* 145 (December 2020). https://doi.org/10.1016/j.envint.2020.106066.

3

PLASTIC IMITATING LIFE

Future Potential for a Controversial Material

Blaine Brownell

Introduction

Plastic is the great ersatz material. It is also the exemplar of unintended consequences. Plastic embodies contradiction and controversy. One of the most capacious, popular, and inexpensive substances has also become one of the most significant threats to human and planetary health. Good and bad, plastic's influence on modern society is profound. Perhaps more than any other material, plastic aims to mimic life. Modern plastic holds a mirror to contemporary human society; it reinforces our tireless search for convenience, preservation, ease, and affordability—as well as our overconsumption, superficiality, and wastefulness. If we consider technology as a means to amplify human potential, plastic is a medium that has served to fulfill our many desires—including protection, preservation, filtration, manipulation, transmission, and communication. Plastic embodies this paradox today: it has become both indispensable and deeply problematic. The material catalyzes human ingenuity while threatening planetary health—and, by extension, our future.

Significantly, plastic is not singular, but encompasses a sweeping collection of synthetic and natural polymers, each with distinctive characteristics, uses, and environmental consequences. This wide variety results in opportunities and challenges. While function-tailored polymers have enabled remarkable advances in lightweight structures, robust apparel, and life-preserving medical devices, the proliferation of complex chemical formulations and multi-material composites lacking a safe means of disposal or reuse has fueled human and environmental health problems. In our growing comprehension of these intricacies, we must emphasize the need to consider plastic as more than a monolithic substance, acknowledging its vast range of capacities and risks.

Plastic is continuously evolving. Since the invention of Parkesine in the mid-nineteenth century, this burgeoning material family has expanded in variety and capability, fulfilling a similarly broad spectrum of human ambitions.[1] Early plastics such as Bakelite and celluloid demonstrated the material's capacity to mimic nature, fueling our yearning for luxury by providing cheap substitutes for scarce natural materials like amber and ivory. Although this copycat aim remains one of plastic's most significant roles (e.g., vinyl siding and flooring

DOI: 10.4324/9781003613589-4

with faux wood grain), the material family has expanded beyond visual and tactile cues to include function and life cycle in its mimetic lexicon. Increasingly, the sophistication and diversity of plastic's capabilities are more closely approximating how life performs—not just how it appears. Responsive, shape-shifting, self-repairing, and multifunctional polymers offer unprecedented biomimicry and biocompatibility. These remarkable substances represent a new epoch for plastic, offering a compelling outlook of a substance increasingly aligned with ecological systems and processes. Because not all new polymers promote environmental health, it is paramount that the continued growth and development of Plastic 2.0 be steered toward ecologically conscientious goals—without undermining its ever-expanding capaciousness.

This essay considers the way forward in three parts: the emerging functional capacities of plastic, the growing knowledge of plastic's harmful effects, and the life cycle-conscious developments that promise to overcome plastic's most significant dangers. Each dimension considers how plastic shapes human society, presaging that a deliberate alteration of its trajectory will transform society. In this way, plastic's continued evolution is inextricably linked to our own.

Functional Advances

If the first generation of plastic prioritized apparent substitution, the new generation finds life-like functional potentials not often found in other human-made materials. In plastic's first era, the material was primarily used skeuomorphically, purposefully emulating the appearance and tactility of other natural substances. This form of mimicry continues to be a common strategy for plastic, particularly whenever the material is employed in visible applications in the built environment. In such cases, plastic almost always operates as a material chameleon—e.g., countertops resembling stone, siding imitating wood, or upholstery and finishes copying natural fabric. When hidden from view, such as PVC piping and conduit, the copycat imperative disappears. But out in the open, plastic has rarely been permitted to be "itself." One reason is that Plastic 1.0 is deemed inferior to the more expensive, less available, and more durable materials it typically emulates, and it represents the built environment's widespread "dirty secret" hidden in plain sight. Another reason is that plastic's essential identity is uncertain, given its near-limitless manifestations and continuous evolution. However, Plastic 2.0's newfound properties, processes, and capabilities invite these novel polymers to be appreciated on their own terms. The following examples represent the new possibilities enabled by human ingenuity, experimentation, and ambition—although the consideration of broader environmental consequences may be absent.

Digital Fabrication

Advanced manufacturing processes are transforming how polymers are shaped into various products, enabling ever more intricate fabrications. 3D printing has increased the potential of achieving complex forms and patterns in a more accessible way than traditional manufacturing methods. The enhanced fabrication opportunities are evident in works like Dutch designer Lilian van Daal's Biomimicry 3D-printed soft seat (Figure 3.1).[2] Inspired by the distinctive patterns of plant cells, the chair features graduated material density—offering greater porosity in areas requiring more flexibility. Fabricated entirely out of polyamide by 3D Systems Benelux,

FIGURE 3.1 Biomimicry (detail) by Lilian van Daal. Photograph by Manon van Daal.

the piece demonstrates the potent connections between mechanical and aesthetic performance, made possible by new techniques.

Multi-material 3D printing enables fabrications with more than one substance, although printers must switch between different printing nozzles. Co-injection molding allows two or more polymers to be added to the same mold simultaneously, enabling a functional differentiation between each material—such as skin and core.[3] A significant drawback to multi-material constructions concerns the lack of recyclability, as it is difficult to separate the original polymers, and many multi-material products wind up in landfills as a result.[4] A new process called Speed-Modulated Ironing, developed by MIT and Delft University of Technology researchers, allows designers to create multi-material effects using a singular feedstock.[5] The method features two nozzles: one to spool heat-responsive filament and the other to manipulate the filament via heat to modulate properties such as texture or opacity. The result is a mono-material that functions like a multi-material construction while maintaining maximum recyclability.

Self-Repair

Enhancing mechanical performance is also a common theme in materials research and development, and polymer scientists are routinely concerned with extending the longevity of plastics in a life-like manner. Polymers exhibiting self-healing properties offer the extraordinary capacity to increase their lifespan after damage. One of the first such examples was developed by the Autonomous Materials Systems Group at the University of Illinois' Beckman Institute for Advanced Science and Technology (Figure 3.2).[6] Inspired by mammals' natural blood-enabled

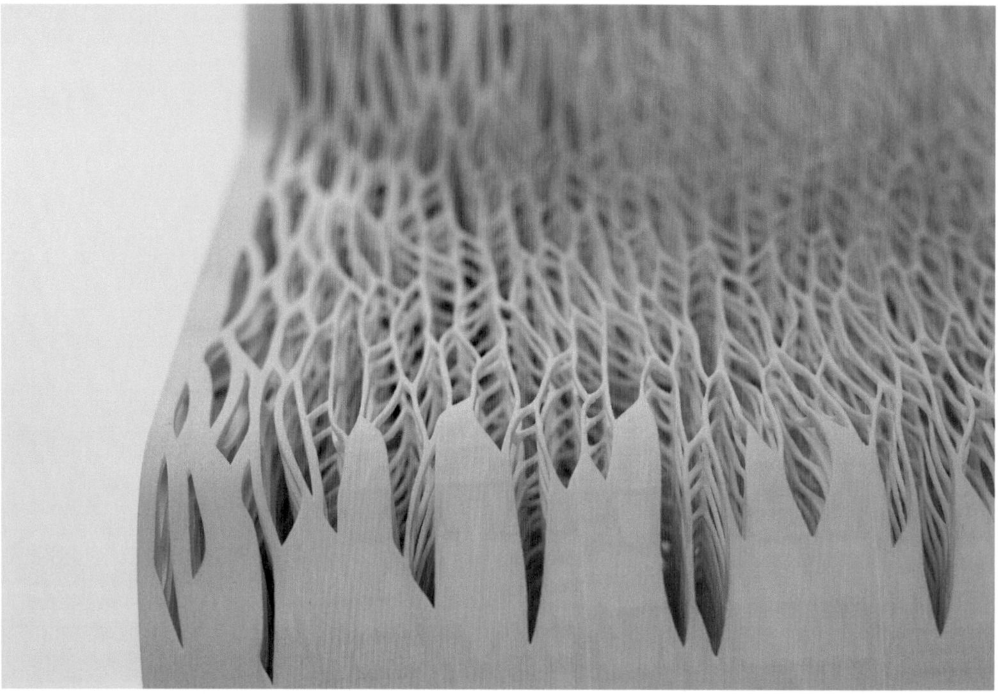

FIGURE 3.2 Self-healing polymer, Beckman Institute, The University of Illinois. Photograph by Magnus Andersson, University of Illinois.

healing capacity, the researchers incorporated microcapsules containing a healing agent within a structural polymer. When cracks form in the material, the microcapsules break open and release the restorative healing agent.

Reducing the duration of the healing process is an aim for polymers in high-risk environments. University of Michigan and NASA scientists collaborated to develop a polymer composite that self-heals instantaneously.[7] The composite consists of three light-transmitting materials: two transparent polymer sheets encapsulating a layer of clear resin. When the materials are punctured, the resin bleeds to the surface, hardening within seconds after contacting atmospheric oxygen. In a compelling demonstration simulating the breach of the window of an airplane or spacecraft, the researchers shot a bullet through the composite.[8] In less than one second, the healing agent sealed the puncture—thus demonstrating potential life-saving capabilities in aerospace applications.

Early self-repairing polymers exhibited a one-time healing process. However, the repeatable self-healing of living organisms is the preferred capacity. The Beckman Institute scientists use the term "living polymerization" to describe this phenomenon, which further extends materials' useful lives.[9] Scientists at the IK4-CIDETEC Research Center developed a remarkable repeatable healing technology by creating a polyurethane urea elastomer (PUU) that self-heals after being cut and wholly divided into distinct parts (Figure 3.3).[10] Before this invention, polymers required an external stimulus such as heat or light for this level of self-restorative capability—however, the Self-Healing PUU functions at room temperature without outside influence. In this case, the PUU's aromatic disulfides initiate a metathesis reaction when in contact, enabling the formation of new bonds between individual pieces.

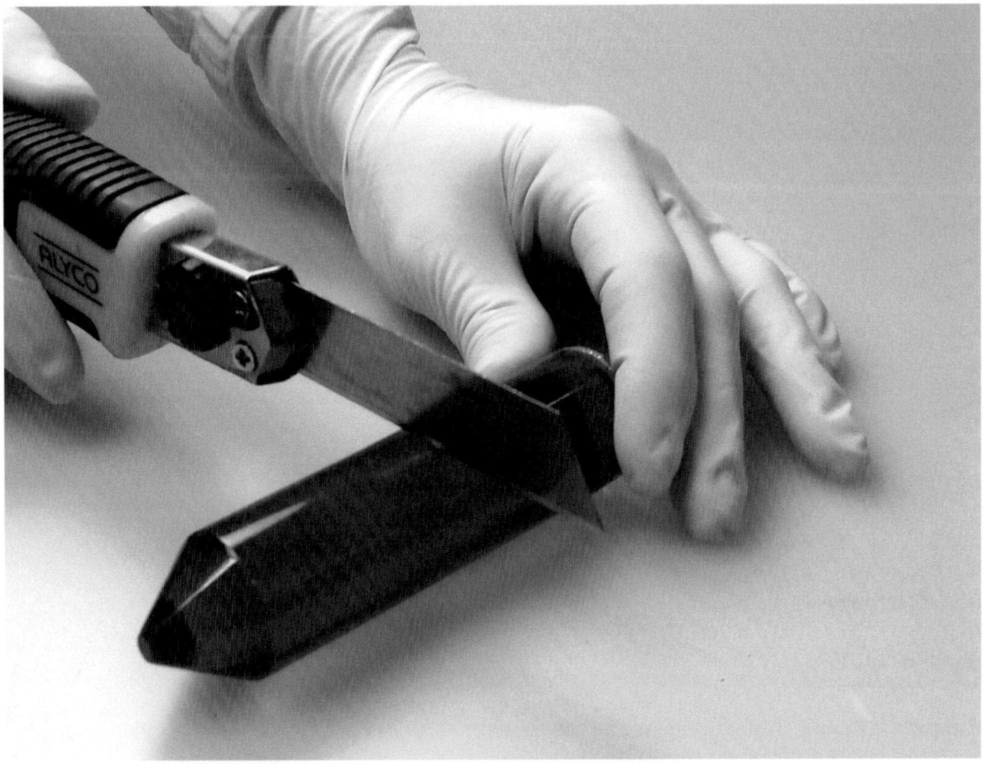

FIGURE 3.3 Self-healing PUU by the IK4-CIDETEC Research Center. Photograph by Roberto Martín.

Mutation

Some novel plastics can transform their physical properties in ways that defy expectations. For example, liquid crystal elastomers (LCEs) are shape-reactive polymers that respond to environmental stimuli counterintuitively (Figure 3.4).[11] When compressed, LCEs don't become weaker; they increase in stiffness. Developed by Rice University scientists from polysiloxane, liquid crystal, and a cross-linking material, LCEs are ideal for structural applications with continuous pressure and increasing loads. The material also functions as a sensor capable of maintaining a memory of past compression events. In addition, the researchers found that they can alter the material's surface texture by adding electrical current. When a voltage is applied, the LCE nanocomposite forms wrinkles—a reversible and repeatable process. Given their biocompatibility, LCEs have been identified as good candidates for tissue scaffolds and other bio-engineering applications, and heart muscle cells have been successfully grown on the wrinkle-stimulated surface.

Researchers at Harvard's Wyss Institute for Biologically Inspired Engineering have developed another mutant material: a polymer that can transform the hydrophobicity of its surface.[12] The elastic substrate, which is perforated with micropores, has a liquid film coating. Manipulations of the surface's shape alter three characteristics—wettability, smoothness, and opacity—based on the deformation of the pore sizes. When at rest, the material is hydrophobic, smooth, and transparent. But when bent or warped, the coating becomes hydrophilic, rough, and

FIGURE 3.4 Liquid crystal elastomer by the Verduzco Laboratory, Rice University. Photograph by Jeff Fitlow/Rice University.

opaque. These changes are immediate, precisely controlled, and infinitely repeatable, as they are caused by simple physical alterations without needing other stimuli. The Wyss researchers were inspired by natural examples like human tears, which exhibit multiple tunable functions like waste removal and protection from foreign particles. Potential applications include fluid-regulating plumbing and pipelines, tunable textiles, and responsive building envelopes.

Light and Communication

Many emerging plastics can harvest, store, emit, and manipulate light. Some polymers can change color, pattern, or texture in a chameleon-like response to external stimuli. Natural phenomena inspiring these developments include luminescence, photonic tuning, and thermochromism. The Yeadon Space Agency has developed novel photoluminescent bioplastics that capture and release contextual light. Architect Peter Yeadon harvests polymers from unconventional streams, such as industrial biowaste, and dopes them with various impurities to increase their illumination capacity. An example polymer mix includes collagen proteins from cows, triglycerides from biodiesel production, and light-emitting colloids. Yeadon has employed photoluminescent polymers to create light-harvesting furniture, such as the BioBench, nocturnal wayfinding-assistive seating designed with designer Danielle Storm Hoogland.[13]

Novel display technologies like electronic ink utilize polymers to function. Like traditional ink, electronic ink consists of colored liquid that can transform a surface into a communicative medium. In the case of e-ink, microcapsules containing two-toned polymer particles are suspended within the fluid.[14] The light and dark sides respond to opposing electrical charges, so an electric field will stimulate the polymers to re-orient themselves according to the types of charges. Although the most familiar e-ink displays are found in portable tablet form, larger displays intended for environmental graphics also exist. For example, E Ink Prism incorporates bistable electronic ink film in modular tiles. The bichromic panels can change between two colors or graduated hues in between with minimal electrical current. E Ink Prism can cover an expansive surface like a wall or ceiling, transforming it into a dynamic communication device that is more integral than a surface-mounted digital display.

Color-changing capabilities are also seen in thermochromic polymers, including some that respond to non-electrical stimuli such as heat. Australia's Commonwealth Scientific and Industrial Research Organization developed a Smart Bandage consisting of a fiber impregnated with thermochromic liquid crystalline (Figure 3.5).[15] As its name suggests, the bandage is designed to assist in monitoring wound healing by changing color according to minute alterations in skin temperature. The polymer fibers' color spectrum ranges from 77 to 113°F and the gradient visibly shifts with local temperature fluctuations as minute as 0.9°F. The ability to communicate infection or other healing problems early, before visible cues might provide such evidence, enables wound dressings to remain undisturbed and, therefore, more effective.

Drawbacks

These increasingly life-like properties demonstrate plastic's remarkable trajectory from a rudimentary ersatz material to a substance of sophisticated, responsive, and self-preserving capabilities akin to living organisms. And yet, these nature-emulating effects are often present without consideration for nature's metabolic processes. This lack of concern for broader ecosystem health results in products that are environmentally obsolete upon their creation (Figure 3.6). The materials may have compelling capacities, but their inability to readily assimilate into industrial ecological flows makes it highly likely they will contribute to the growing polymer refuse stream. For example, self-healing plastics embody impressive restorative capacities to extend functional viability and longevity. However, when composed of non-biocompatible substances, these polymers diminish the health of the larger ecosystems they ultimately permeate. Thus, two natural objectives—endurance and vitality—are in direct conflict. This contradiction illuminates a recurring characteristic of human behavior, an anchoring cognitive bias toward a singular object, and in its present condition, at the expense of the broader spatiotemporal context.

In this way, plastic's worst offenses concern its unintended consequences—ramifications resulting from the multiple forms of waste it produces. The most visible kind of waste is the plastic rubbish clogging the world's waterways and landscapes (Figure 3.7). Less-visible sources of waste include the toxic, pervasive chemicals and microparticles now present on every continent and in most living organisms. There is also the waste of greenhouse gases emitted when virgin plastic is created from petrochemicals. None of these insights is unfamiliar, and this book is undoubtedly replete with a thorough accounting of plastic's devastating effects. For this reason, a comprehensive scientific accounting is not included here. Instead, plastic's primary

FIGURE 3.5 Smart Bandage. Photograph by CSIRO.

FIGURE 3.6 A monkey inspects a disposable plastic bag in Kathmandu, Nepal. Photograph by Davis Brownell.

drawbacks are addressed via three material stories, each offering a concrete example of plastic's most egregious offenses and their relationships to human behavior.

Hypertoxicity

On February 3, 2023, a 38-car-long Norfolk Southern train derailed in East Palestine, Ohio. Carrying hazardous chemicals such as vinyl chloride, the highly flammable and carcinogenic gas used to create polyvinyl chloride (PVC), the train exploded and burned for two days.[16] According to the National Cancer Institute, vinyl chloride exposure increases the risk of cancers of the liver, lungs, and brain, including lymphoma and leukemia.[17] Based on the knowledge of these harrowing effects, public officials ordered residents living within a one-mile radius of the crash to evacuate.

The East Palestine disaster dominated news headlines for days, and the bulk of criticism was directed toward Norfolk Southern and the US Department of Transportation for failing to prevent the accident. However, any criticism of PVC or the industries that rely upon it was conspicuously absent. Vinyl is the most used plastic in building construction, with about 70 percent of all PVC directed toward building applications.[18] The material's commercial demand is well-established,

FIGURE 3.7 Discarded beverage containers fill a public fountain in Astana, Kazakhstan. Photograph by the author.

with a market value of over $80 billion in 2023.[19] Perhaps it is for this reason that none of the major news outlets questioned society's reliance upon such a hazardous substance. Other likely possibilities are the convenience and affordability that vinyl brings to our contemporary way of life. And yet, despite the hyper-toxicity of PVC's chemical origins, the fact that there are viable alternatives to vinyl in construction, and the knowledge that accidents are unavoidable (1000 train derailments occur in the United States annually), society is strangely comfortable choosing this myopic and self-detrimental path.[20]

Ubiquity

Plastic is inescapable. In the relatively brief period since the invention of synthetic polymers, they have expanded their physical presence into every corner of the planet. The effects of plastic's omnipresence are not entirely known, but what evidence does exist is alarming. The latest concern is the proliferation of microplastics—tiny polymer fragments between 1 nm and 5 mm in size—from both new and discarded products.[21]

In December 2022, University of Auckland scientists published their finding that no fewer than 74 metric tons of microplastics descend from the atmosphere above the city each year.[22]

This quantity, estimated to occupy nearly 5,000 m^2 of area per day, is equivalent to more than three million plastic beverage containers. The researchers identified eight common polymer types from these particles, all of which are petroleum-derived, including polyethylene (PE), polystyrene (PS), polypropylene (PP), PVC, and polycarbonate (PC). Although Auckland was the focus of this study, it is believed that a high quantity of microplastics exists everywhere.

Early studies of microplastics' effects are disturbing. A comprehensive study investigating the particles' toxicological impacts on human cells found several types of damage, including cell wall deterioration, allergic reactions, and cell death.[23] Furthermore, the degradation of plastic particles functions to increase their danger—not decrease it as in most other materials. As the particles are further fragmented from micro to nano size, their distribution area expands, and they infiltrate the cells of smaller, more primitive organisms at the base of the food chain. Nanoplastics are considered more harmful because they can penetrate biological membranes.[24] As Hull York Medical School microplastics researcher Evangelos Danopoulos warned *The Guardian*, "We should be concerned. Right now, there isn't really a way to protect ourselves."[25]

Demand

One might presume that the growing awareness of plastic's detrimental influence would make it a less popular material, or that the increased capability to recycle existing polymers effectively would taper the growth of virgin petroleum-based plastics. Neither is the case. The International Energy Agency (IEA) has determined that petrochemicals—which include plastics—are growing faster than any other bulk material.[26] Demand for industrial polymers grew tenfold between 1971 and 2015, more than cement, aluminum, or steel. And petrochemicals' surge is anticipated to outpace the growth of oil. "Petrochemicals will be the largest driver of global oil demand over the coming decades, ahead of cars, trucks, and planes," the IEA declared.[27] "They will account for more than a third of the growth in oil demand to 2020 and nearly half the growth to 2050." This meteoric rise is concerning because petrochemicals will eventually be the most significant product from—and contributor to—fossil fuel production, adding an estimated daily consumption of seven million barrels of oil by 2050. This output amounts to over three million metric tons of CO_2 emitted daily, or one billion metric tons annually, for petrochemicals alone. As humanity's demand for materials continues to accelerate, it hungers for this resource above all.

Life Cycle Advances

Despite the bad news about plastic, many compelling new polymers have been developed to address these and other environmental challenges. In direct response to the most egregious ills mentioned above, examples include fully biocompatible and nontoxic plastics, plastics designed to biodegrade safely at the end of their useful lives, and plastics that utilize feedstocks other than petrochemicals. Of additional note are polymer technologies developed to reduce embodied and operational energy, sequester carbon, and find new uses for discarded plastics. Although bioplastic processing remains energy intensive, and more research is required to understand the effects of biobased microplastics, these technologies offer significant advantages over their traditional counterparts. Most importantly, these environmentally responsive next-generation polymers are designed with ecosystem health as a priority.

Biobased Feedstocks

Material scientists seeking to avoid fossil fuels tend to gravitate toward a biological source of the original plastic feedstock: cellulose. The most common organic polymer on the planet and the fundamental structure of plant cell walls, cellulose formed the basis for Parkesine, the original thermoplastic developed by Alexander Parkes in the mid-nineteenth century from nitrocellulose and camphor.[28] Recently, Australia-based Zeoform devised a method of extracting cellulose from reclaimed biobased materials, including textiles, paper, and hemp, to create a new plastic (Figure 3.8).[29] The Zeoform polymer derives its robustness from hydroxyl bonding, the same fiber-strengthening process found in plants. The biobased plastic consists of only two ingredients—cellulose and water—and contains no binders or additives.

Cellulose is also a viable feedstock for 3D printing. Swedish scientists have developed variants of 3D-printable nanocellulose, a version with high-strength nanoscale fibers. Umeå University researchers have 3D-printed experimental building components, from weatherstripping to walls, from nanocellulose as part of the so-called +Project.[30] A collaboration between Sweden's Chalmers University of Technology and the Wallenberg Wood Science Center resulted in robotically fabricated architectural membranes out of nanocellulose hydrogel, a pliable medium with increased workability due to the addition of alginate.[31] Notably, the nanocellulose printing process is conducted at room temperature. No heating is required; the only processing energy involves pressurizing the air needed to maintain the liquid state of the feedstock until it is deposited.

Another alternative feedstock gaining attention is chitin. The second most prevalent natural polymer after cellulose, chitin forms the cell walls of other organisms, including insects, crustaceans, and fungi. Researchers at Harvard's Wyss Institute for Biologically Inspired

FIGURE 3.8 Zeoform cellulose-based polymer by Zeoform. Photograph by Zeoform.

Engineering have developed a biopolymer from chitosan, a deacetylated and more soluble variant of chitin, and wood flour.[32] The scientists processed the chitin from discarded shrimp shells and created a thin, light-transmitting film that exhibits the strength of aluminum at only half its weight. Optimal for small-scale applications ranging from medical supplies to mobile electronics, the chitosan bioplastic is a biocompatible, biodegradable, waste-reducing alternative to traditional petroleum-based plastics.

Biocomposites

One of the most intriguing developments in building construction applications is biocomposite envelope systems. Scientists and manufacturers are creating biobased versions of facade components that are typically composed of petroleum-based plastics and concrete, metal, and wood. The European Commission-sponsored BioBuild project is reportedly the first self-supporting biocomposite facade panel.[33] Developed by Arup and GXN Innovation, the natural fiber and resin-based panel consists of two outer bio-polyester-reinforced flax fiber shells incorporating a window. Each 13' × 7'-6" (4 m × 2.3 m) module represents half the embodied energy of a conventional facade system of the same area. The BioBuild project also resulted in three other building systems, including a rain-screen assembly consisting of jute and furan resin, interior partitions made of flax and furan, and acoustic ceiling tiles composed of jute and bio-polyester.[34]

UNStudio and 17 other organizations also developed new biocomposite assemblies as part of European Union-supported research and development called the Osirys Project.[35] Each assembly features biocompatible material replacements for most of the common elements of a conventional version. For example, the curtain wall system replaces the aluminum frames and profiles with biocomposite profiles, covers, and shading panels. Like BioBuild, the project resulted in multiple envelope and interior partition systems. Osirys components and assemblies have low carbon footprints and contribute fewer volatile organic compounds (VOCs) than the conventional systems they replace, resulting in cleaner air.

Fiber producer Bcomp collaborated with composite manufacturer Temca to develop a novel biobased trellis for the facade of the Cité Scolaire Internationale Jacques Chirac school in Marseille, France.[36] The latticework helps shade the building glazing while providing a visually striking pattern for the facade. Each panel is made from flax fibers reinforced with glass fibers and polyester resin in one of five different molds. When the panels are combined to form a seamless field, the variety of panel geometries minimizes visible module repetition. The tracery is treated with a UV-protective coating for enhanced durability. The Bcomp/Temca trellis extends to buildings the concept of lightweighting, a strategy to decarbonize the automobile industry with lighter-weight and lower-impact components.[37] "To our knowledge, this is the first architectural project of this size featuring natural fibres," said Paolo Dassi, Bcomp Industry Manager. "Sustainable lightweighting is not just for automotive vehicles!"[38]

Carbon Sequestration

Wood has received the most attention as the building material that supports carbon drawdown based on its intrinsic carbon storage before harvesting. However, plastics can also contribute to the goal of atmospheric carbon reduction, and several polymers are now manufactured using carbon dioxide as a raw ingredient. For example, Aircarbon is a novel plastic made from greenhouse gas emissions.[39] Huntington Beach, California-based Newlight Technologies converts CO_2 and methane into polymers for a wide variety of commercial-scale applications. The process begins by growing microorganisms in salt water, which absorb these greenhouse gases and convert them into long-chain polymers.[40] The company isolates the polymer, filters it, and purifies it before extruding it into strands that it cuts into pellets. According to certified third-party tests, Aircarbon removes more carbon from the atmosphere than it contributes.

Rochester, New York-based Novomer—now part of Danimer Scientific—also manufactures polymers from greenhouse gases.[41] The company uses plant-based, renewable ingredients to create low-carbon polymers and chemicals via its Novo 22 Catalyst and conversion process. Novomer utilizes two biobased feedstocks: one from biomass and biobased waste streams and one from sugar. Rinnovo, the manufacturer's primary product, is designed to be readily composted or recycled at the end of its life.[42] In addition, the polymer may be upcycled via thermolysis to form acrylic acid, the raw chemical ingredient for various high-performance chemicals. Novomer aims to complete the carbon cycle in a simple and direct way: "While many circular economy diagrams involve many 'gears,' our impact concept is simple: start with sustainable feedstock, convert it as efficiently as possible and return it back to its original form to start the process over."[43]

CO_2 may also be used in place of conventional solvents used in coatings and paints. Researchers at the Fraunhofer Institute for Environmental, Safety, and Energy Technology (UMSICHT) in Oberhausen, Germany, developed a method to impregnate polymers like polycarbonate and nylon with CO_2 (Figure 3.9).[44] The process begins by heating CO_2 to 30.1°C under pressure, which makes it behave like a solvent. Increasing the pressure further enables powdered pigment to be dissolved within the converted substance, allowing it to function like a colorant for plastics. Other potential functions of the additive include imparting antimicrobial and heat-responsive characteristics.

Second Life

Human society generates over 350 million metric tons of plastic waste each year.[45] The proliferation of this non-biocompatible substance worldwide will eventually produce significant quantities of micro- and nano-plastics over the coming centuries. Recycling is no longer an option but a necessary strategy to redirect waste into new products while reducing humanity's reliance upon virgin feedstocks. This is the mission of the Good Plastic Company, an Amsterdam-based manufacturer of second-life polymer products, and the growing number of others like it.[46] The company's Polygood panels are made from recycled PS, a typical consumer polymer with good mechanical properties. Old refrigerators, televisions, yogurt containers, and many other PS-based consumer waste items comprise 70 percent of the panels, while the remaining 30 percent consists of post-industrial refuse like PS spools and tubing. Polygood panels are made via thermopressing, which requires less energy than conventional molding methods, and uses no additives, coatings, or fillers. The Good Plastic Company offers a take-back program to maintain a closed loop for this inherently recyclable material.

Contrasting with the refined and feedstock-conscientious process to make Polygood panels is ByFusion's indiscriminate approach to creating raw building components from plastic trash.[47] The Los Angeles-based company's compact processing technology, the Blocker, is designed to be transported via shipping container to waste dumps. The assembly shreds waste plastic into smaller fragments, melts these via an integral water boiler, and presses the resulting liquid into concrete masonry unit-sized construction modules. These raw building blocks have

FIGURE 3.9 CO_2-impregnated polymers by the Fraunhofer Institute for Environmental, Safety, and Energy Technology UMSICHT. Photograph by Fraunhofer UMSICHT.

higher compression strength, contribute 83 percent fewer CO_2 emissions than non-load-bearing CMUs, and are suitable for constructions like accent walls, privacy screens, and utility sheds.

Puebla, Mexico-based EcoDom Brikstudio offers larger-scale building assemblies made from plastic waste.[48] The company's Thermal Wall is an 8' × 2' insulated structural panel of repurposed plastic trash compressed into a steel rebar frame. The plastic offers acoustic and thermal insulation, while the rebar provides reinforcing against gravity and seismic loads. EcoDom's Thermal Roofing modules are 4' x 4' plastic-clad, precast concrete roofing panels. A 1-inch thick layer of concrete functions as the visible ceiling plane, above which lies a thermal barrier of recycled waste plastic and a steel rebar and wire mesh structural frame. The company advertises the capability to recycle 15,000 kg of solid waste—which translates to 1,200 load-supporting walls, roofs, and floors—every week, representing a significant endeavor to redirect plastic trash into viable elements for new residential architecture.

The Way Forward

Bioplastics only account for 0.5 percent of the 400 million metric tons of polymers in current production.[49] However, bioplastics are growing at a significant pace—from over two million metric tons in 2023 to an expected 7.4 million metric tons in 2028.[50] The majority of this quantity will be biodegradable polymers, at 4.6 million metric tons. While most bioplastics are used for low-value, temporary uses like flexible packaging, more biopolymers are appearing in high-value, durable applications like automobiles and building construction. Furthermore, the demand for recycled plastics is also increasing. In 2022, recycled polymers comprised only about nine percent of global plastic production.[51] Nevertheless, the recycled plastics market is anticipated to nearly double between 2023 and 2030, from $69 to $120 billion.[52]

Realistically, these alternative polymers have a long way to go to catch up to their petroleum-based counterparts—even though future planetary health will be influenced by this transition. These alternative material streams also require more scrutiny to validate their environmental merits. For example, bioplastics may have a reduced carbon footprint, but their environmental impact remains high. Biopolymers typically rely on large quantities of land area and chemical fertilizers, resulting in higher aquatic eutrophication and soil acidification than conventional plastics. Bioplastics also rely too heavily on edible feedstocks like corn and sugar, reinforcing critics' arguments that they exploit precious arable land needed for food. Biodegradability and compostability are also casually used terms that do not convey significant details, such as the time or conditions required for complete bio-assimilation or the potential for adverse environmental impacts when discarded in an inappropriate context. Additionally, microplastics remain an issue that must be addressed with these alternative polymers as well as traditional ones. All of these challenges must be addressed in further biopolymer development.

In *A New Reality: Human Evolution for a Sustainable Future*, the late scientist Jonas Salk and his son Jonathan Salk project humanity's possible futures based on biological models of organisms in finite containers not unlike our own.[53] Population growth always follows an S-curve, accelerating at the beginning until it hits an inflection point as it approaches resource limits. Afterward, the community's careful use of resources will allow it to continue, although the growth slows. Humanity is now encountering this inflection point in its own growth on this finite planet, and how we use or misuse resources will determine our future trajectory beyond this threshold.

Plastic is a material mirror of modern society, reflecting our aspirations for convenience, technological advancement, and natural connections. Plastic's ties to nature began with aesthetic emulation and have recently developed two additional forms of imitation: functional performance and material life cycle. While traditional plastic's downsides are profoundly detrimental, aligning with living systems can transform plastic into a wholly beneficial material. We must, therefore, find ways that plastic can continue to advance in functional performance as a holistically biocompatible resource. This is a second chance for plastic to fulfill its original vision without detrimental effects. So, as plastic continually seeks to imitate life, it must eventually reconcile with it. The alternative path is unsustainable.

Notes

1 Seth C. Rasmussen, "From Parkesine to Celluloid: The Birth of Organic Plastics," *Angewandte Chemie (International ed. in English)* 60, no. 15 (2021), https://doi.org/10.1002/anie.202015095.
2 "Biomimicry: 3D Printed Soft Seat," Lilian van Daal Studio Lilian van Daal, accessed January 10, 2025, www.lilianvandaal.com/biomimicry3dprintedsoftseat.
3 Wenjun He et al., "Study on Co-Injection Molding of Poly(Styrene-Ethylene-Butylene-Styrene) And Polypropylene: Simulation and Experiment," *Polymer Testing* 108 (2022), https://doi.org/10.1016/j.polymertesting.2022.107510.
4 "Multi-Layer Plastic Packaging: Recycling Challenges and Perspectives," Fabian Arevalo Plastics Engineering, accessed January 10, 2025, www.plasticsengineering.org/2024/05/multi-layer-plastic-packaging-recycling-challenges-and-perspectives-004634/.

5 "New 3D Printing Technique Creates Unique Objects Quickly and with Less Waste," Adam Zewe MIT News, accessed January 10, 2025, news.mit.edu/2024/new-3d-printing-technique-creates-unique-objects-using-less-waste-1010.
6 "Autonomic Materials Group," Autonomic Materials Group, Beckman Institute for Advanced Science and Technology, accessed October 6, 2025, http://autonomic.beckman.illinois.edu/.
7 Scott R. Zavada et al., "Rapid, Puncture-Initiated Healing via Oxygen-Mediated Polymerization," *ACS macro letters* 4, no. 8 (2015), https://doi.org/10.1021/acsmacrolett.5b00315.
8 "'Terminator'-Style Material Heals Itself," American Chemical Society, accessed January 10, 2025, www.youtube.com/watch?v=JVWFvKxrcLg.
9 "Microencapsulation Approaches," Autonomic Materials Group, Beckman Institute for Advanced Science and Technology, accessed October 6, 2025, http://autonomic.beckman.illinois.edu/microencapsulation.html.
10 "Self Healing Polymer by CIDETEC," CIDETEC, accessed January 10, 2025, www.youtube.com/watch?v=fVTfSHEPnr8.
11 "Liquid Crystal Elastomers," Verduzco Laboratory Rice University, accessed January 10, 2025, https://verduzcolab.blogs.rice.edu/liquid-crystalline-order/.
12 "Cry Me a River of Possibility: Scientists Design New Adaptive Material Inspired by Tears," Wyss Institute for Biologically Inspired Engineering at Harvard University, accessed January 10, 2025, https://wyss.harvard.edu/news/cry-me-a-river-of-possibility-scientists-design-new-adaptive-material-inspired-by-tears/.
13 "Yeadon Space Agency Helps Storm Hoogland Create BioBench … Landscape Furniture That's Not Meant to Last," Yeadon Space Agency Yeadon Space Agency, accessed January 10, 2025, www.yeadonspaceagency.com/news/.
14 "Electronic Ink," How Products Are Made, accessed January 10, 2025, www.madehow.com/Volume-6/Electronic-Ink.html.
15 "Smart Bandages Can Help Identify Infection," ATA Textile Technology Source, accessed January 10, 2025, https://textiletechsource.com/2017/02/24/smart-bandages-can-help-identify-infection/.
16 Christine Hauser, "After the Ohio Train Derailment: Evacuations, Toxic Chemicals and Water Worries," *The New York Times*, 2023, accessed January 10, 2025, www.nytimes.com/article/ohio-train-derailment.html.
17 "Vinyl Chloride," National Cancer Institute at the National Institutes of Health U.S. National Cancer Institute, accessed January 10, 2025, www.cancer.gov/about-cancer/causes-prevention/risk/substances/vinyl-chloride.
18 "PVC Materials in Building, Construction & Design," Vinyl Institute, accessed January 10, 2025, www.vinylinfo.org/uses/building-and-construction/.
19 "$120+ Billion Polyvinyl Chloride (PVC) Market Opportunities and Strategies to 2033: Expansion of Production Capacities to Cater to the Increased Demand Across Diverse Industries," Research and Markets, accessed January 10, 2025, www.globenewswire.com/news-release/2024/11/28/2988784/28124/en/120-Billion-Polyvinyl-Chloride-PVC-Market-Opportunities-and-Strategies-to-2033-Expansion-of-Production-Capacities-to-Cater-to-the-Increased-Demand-Across-Diverse-Industries.html.
20 Luz Lazo, "Derailments Aren't Uncommon, but Hazmat Spills Are Rare. Here's What We Know," *The Washington Post*, 2023, accessed January 10, 2025, www.washingtonpost.com/transportation/2023/02/17/train-derailment-statistics-safety-damages/.
21 "Microplastics," Illinois Environmental Protection Agency Illinois Environmental Protection Agency, accessed January 10, 2025, https://epa.illinois.gov/topics/water-quality/microplastics.html.
22 "Microplastics in Auckland's Air Revealed," University of Auckland Waipapa Taumata Rau, University of Auckland, accessed January 10, 2025, www.auckland.ac.nz/en/news/2022/12/12/microplastics-in-air.html.
23 Damian Carrington, "Microplastics Cause Damage to Human Cells, Study Shows," *The Guardian*, 2021, accessed January 10, 2025, www.theguardian.com/environment/2021/dec/08/microplastics-damage-human-cells-study-plastic.
24 Maxine S.-L. Yee et al., "Impact of Microplastics and Nanoplastics on Human Health," *Nanomaterials (Basel, Switzerland)* 11, no. 2 (2021), https://doi.org/10.3390/nano11020496.
25 Carrington, "Microplastics cause damage to human cells, study shows."
26 International Energy Agency, "The Future of Petrochemicals: Towards more sustainable plastics and fertilisers" (OECD/IEA, Paris, 2018), https://iea.blob.core.windows.net/assets/bee4ef3a-8876-4566-98cf-7a130c013805/The_Future_of_Petrochemicals.pdf.

27 "Petrochemicals Set to Be the Largest Driver of World Oil Demand, Latest IEA Analysis Finds," International Energy Agency, accessed January 10, 2025, www.iea.org/news/petrochemicals-set-to-be-the-largest-driver-of-world-oil-demand-latest-iea-analysis-finds.
28 "Alexander Parkes," Chad Peterson, accessed January 10, 2025, https://plasticshof.org/members/alexander-parkes/.
29 "Zeoform," Zeo Sustainable Materials Pty Ltd, accessed January 10, 2025, https://zeoform.com/.
30 "Sliperiet at Umeå University Is Starting a Project: 3D Printing Houses," Umeå University, accessed January 10, 2025, www.umu.se/en/news/sliperiet-at-umea-university-is-starting-a-project-3d-printing-houses_5843377/.
31 "3D Printed Nanocellulose Upscaled for Green Architectural Applications," Chalmers University of Technology, accessed January 10, 2025, www.chalmers.se/en/current/news/3d-printed-nanocellulose-upscaled-for-green-architectural-applications,c3922527/.
32 "Chitosan Bioplastic," Wyss Institute for Biologically Inspired Engineering at Harvard University, accessed January 10, 2025, https://wyss.harvard.edu/media-post/chitosan-bioplastic/.
33 "The World's First Self-Supporting Façade Panel Made of Biocomposites," Arup, accessed January 10, 2025, www.arup.com/en-us/projects/biobuild-facade-system/.
34 "BioBuild," accessed January 10, 2025, https://gxn.3xn.com/project/biobuild.
35 "Forest Based Composites for Façades and Interior Partitions to Improve Indoor Air Quality in New Builds and Restoration," Acciona, accessed January 10, 2025, www.acciona.com/projects/osirys.
36 "Bcomp's Flax Composite Used by Temca to Create Cladding for Shading an International School Building," Bcomp, accessed January 10, 2025, www.bcomp.com/news/bcomp-flax-composite-temca-school-cladding/.
37 "How Can Sustainable Lightweighting Help Decarbonise the Automotive Industry?," Bcomp, accessed January 10, 2025, www.bcomp.com/news/how-can-sustainable-lightweighting-help-decarbonise-the-automotive-industry/.
38 Bcomp, "Bcomp's Flax Composite Used by Temca to Create Cladding for Shading an International School Building."
39 "AirCarbon," AirCarbon, accessed January 10, 2025, www.aircarbon.com/.
40 "Tech," AirCarbon, accessed January 10, 2025, www.aircarbon.com/tech.
41 "Danimer Scientific," Danimer Scientific, accessed January 10, 2025, https://danimerscientific.com/.
42 Danimer Scientific, "Danimer Scientific."
43 Danimer Scientific, "Danimer Scientific."
44 "Impregnation," Fraunhofer UMSICHT, accessed October 10, 2025, www.umsicht.fraunhofer.de/en/carbonmanagement/impregnation.html.
45 "Plastic Waste Worldwide – Statistics & Facts," Statista Statista Research Department, accessed January 10, 2025, www.statista.com/topics/5401/global-plastic-waste/#topicOverview.
46 "The Good Plastic Company," The Good Plastic Company, accessed January 10, 2025, https://thegoodplasticcompany.com/.
47 "ByFusion," ByFusion, accessed January 10, 2025, https://byfusion.com.
48 "Ecodom," Ecodom, accessed January 10, 2025, www.ecodom.mx.brikstudio.com.
49 European Bioplastics e.V., "Bioplastics Market Development Update 2023" (European Bioplastics e.V., Berlin/Brussels, 2023), https://docs.european-bioplastics.org/publications/market_data/2023/EUBP_Market_Data_Report_2023.pdf.
50 European Bioplastics e.V., "Bioplastics Market Development Update 2023."
51 "Plastics Europe Launches Plastics – the Fast Facts 2023," Plastics Europe, accessed January 10, 2025, https://plasticseurope.org/media/plastics-europe-launches-the-plastics-the-fast-facts-2023/.
52 "Recycled Plastic Market," MarketsandMarkets, accessed January 10, 2025, www.marketsandmarkets.com/Market-Reports/recycled-plastic-market-115486722.html.
53 Jonas Salk and Jonathan Salk, *A New Reality*: *Human Evolution for a Sustainable Future*, with the assistance of Jonathan Salk, David Dewane, and Elizabeth H. Blackburn (City Point Press, 2018).

References

Acciona. "Forest Based Composites for Façades and Interior Partitions to Improve Indoor Air Quality in New Builds and Restoration." Accessed January 10, 2025. www.acciona.com/projects/osirys.

AirCarbon. "AirCarbon." Accessed January 10, 2025. www.aircarbon.com/.

AirCarbon. "Tech." Accessed January 10, 2025. www.aircarbon.com/tech.

American Chemical Society. "'Terminator'-Style Material Heals Itself." Accessed January 10, 2025. www.youtube.com/watch?v=JVWFvKxrcLg.

Arevalo, Fabian. "Multi-Layer Plastic Packaging: Recycling Challenges and Perspectives." Accessed January 10, 2025. www.plasticsengineering.org/2024/05/multi-layer-plastic-packaging-recycling-challenges-and-perspectives-004634/.

Arup. "The World's First Self-Supporting Façade Panel Made of Biocomposites." Accessed January 10, 2025. www.arup.com/en-us/projects/biobuild-facade-system/.

ATA. "Smart Bandages Can Help Identify Infection." Accessed January 10, 2025. https://textiletechsource.com/2017/02/24/smart-bandages-can-help-identify-infection/.

Autonomic Materials Group, Beckman Institute for Advanced Science, and Technology. "Autonomic Materials Group." Accessed October 6, 2025. http://autonomic.beckman.illinois.edu/.

Autonomic Materials Group, Beckman Institute for Advanced Science, and Technology. "Microencapsulation Approaches." Accessed October 6, 2025. http://autonomic.beckman.illinois.edu/microencapsulation.html.

Bcomp. "Bcomp's Flax Composite Used by Temca to Create Cladding for Shading an International School Building." Accessed January 10, 2025. www.bcomp.com/news/bcomp-flax-composite-temca-school-cladding/.

Bcomp. "How Can Sustainable Lightweighting Help Decarbonise the Automotive Industry?." Accessed January 10, 2025. www.bcomp.com/news/how-can-sustainable-lightweighting-help-decarbonise-the-automotive-industry/.

ByFusion. "ByFusion." Accessed January 10, 2025. https://byfusion.com.

Carrington, Damian. "Microplastics Cause Damage to Human Cells, Study Shows." *The Guardian*, 2021. Accessed January 10, 2025. www.theguardian.com/environment/2021/dec/08/microplastics-damage-human-cells-study-plastic.

Chalmers University of Technology. "3D Printed Nanocellulose Upscaled for Green Architectural Applications." Accessed January 10, 2025. www.chalmers.se/en/current/news/3d-printed-nanocellulose-upscaled-for-green-architectural-applications,c3922527/.

CIDETEC. "Self Healing Polymer by CIDETEC." Accessed January 10, 2025. www.youtube.com/watch?v=fVTfSHEPnr8.

Danimer Scientific. "Danimer Scientific." Accessed January 10, 2025. https://danimerscientific.com/.

Ecodom. "Ecodom." Accessed January 10, 2025. www.ecodom.mx.brikstudio.com.

European Bioplastics e.V. "Bioplastics Market Development Update 2023." European Bioplastics e.V., Berlin/Brussels, 2023. https://docs.european-bioplastics.org/publications/market_data/2023/EUBP_Market_Data_Report_2023.pdf.

Fraunhofer UMSICHT. "Impregnation." Accessed October 10, 2025. www.umsicht.fraunhofer.de/en/carbonmanagement/impregnation.html.

GXN Innovation. "BioBuild." Accessed February 20, 2026. https://gxn.3xn.com/project/biobuild.

Hauser, Christine. "After the Ohio Train Derailment: Evacuations, Toxic Chemicals and Water Worries." *The New York Times*, 2023. Accessed January 10, 2025. www.nytimes.com/article/ohio-train-derailment.html.

He, Wenjun, Jian Yang, and Yaowei Chen et al. "Study on Co-Injection Molding of Poly(Styrene-Ethylene-Butylene-Styrene) and Polypropylene: Simulation and Experiment." *Polymer Testing* 108 (2022): 107510. https://doi.org/10.1016/j.polymertesting.2022.107510.

How Products Are Made. "Electronic Ink." Accessed January 10, 2025. www.madehow.com/Volume-6/Electronic-Ink.html.

Illinois Environmental Protection Agency. "Microplastics." Accessed January 10, 2025. https://epa.illinois.gov/topics/water-quality/microplastics.html.

International Energy Agency. "*The Future of Petrochemicals: Towards More Sustainable Plastics and Fertilisers*." OECD/IEA, Paris, 2018. https://iea.blob.core.windows.net/assets/bee4ef3a-8876-4566-98cf-7a130c013805/The_Future_of_Petrochemicals.pdf.

International Energy Agency. "Petrochemicals Set to Be the Largest Driver of World Oil Demand, Latest IEA Analysis Finds." Accessed January 10, 2025. www.iea.org/news/petrochemicals-set-to-be-the-largest-driver-of-world-oil-demand-latest-iea-analysis-finds.

Lazo, Luz. "Derailments Aren't Uncommon, but Hazmat Spills Are Rare. Here's What We Know." *The Washington Post,* 2023. Accessed January 10, 2025. www.washingtonpost.com/transportation/2023/02/17/train-derailment-statistics-safety-damages/.

MarketsandMarkets. "Recycled Plastic Market." Accessed January 10, 2025. www.marketsandmarkets.com/Market-Reports/recycled-plastic-market-115486722.html.

National Cancer Institute at the National Institutes of Health. "Vinyl Chloride." Accessed January 10, 2025. www.cancer.gov/about-cancer/causes-prevention/risk/substances/vinyl-chloride.

Peterson, Chad. "Alexander Parkes." Accessed January 10, 2025. https://plasticshof.org/members/alexander-parkes/.

Plastics Europe. "Plastics Europe Launches Plastics – The Fast Facts 2023." Accessed January 10, 2025. https://plasticseurope.org/media/plastics-europe-launches-the-plastics-the-fast-facts-2023/.

Rasmussen, Seth C. "From Parkesine to Celluloid: The Birth of Organic Plastics." *Angewandte Chemie (International ed. in English)* 60, no. 15 (2021): 8012–16. https://doi.org/10.1002/anie.202015095.

Research and Markets. "$120+ Billion Polyvinyl Chloride (PVC) Market Opportunities and Strategies to 2033: Expansion of Production Capacities to Cater to the Increased Demand Across Diverse Industries." Accessed January 10, 2025. www.globenewswire.com/news-release/2024/11/28/2988784/28124/en/120-Billion-Polyvinyl-Chloride-PVC-Market-Opportunities-and-Strategies-to-2033-Expansion-of-Production-Capacities-to-Cater-to-the-Increased-Demand-Across-Diverse-Industries.html.

Salk, Jonas, and Jonathan Salk. *A New Reality: Human Evolution for a Sustainable Future*. With the assistance of Jonathan Salk, David Dewane, and Elizabeth H. Blackburn. City Point Press, 2018.

Statista. "Plastic Waste Worldwide – Statistics & Facts." Accessed January 10, 2025. www.statista.com/topics/5401/global-plastic-waste/#topicOverview.

The Good Plastic Company. "The Good Plastic Company." Accessed January 10, 2025. https://thegoodplasticcompany.com/.

Umeå University. "Sliperiet at Umeå University Is Starting a Project: 3D Printing Houses." Accessed January 10, 2025. www.umu.se/en/news/sliperiet-at-umea-university-is-starting-a-project-3d-printing-houses_5843377/.

University of Auckland. "Microplastics in Auckland's Air Revealed." Accessed January 10, 2025. www.auckland.ac.nz/en/news/2022/12/12/microplastics-in-air.html.

van Daal, Lilian. "Biomimicry: 3D Printed Soft Seat." Accessed January 10, 2025. www.lilianvandaal.com/biomimicry3dprintedsoftseat.

Verduzco Laboratory. "Liquid Crystal Elastomers." Accessed January 10, 2025. https://verduzcolab.blogs.rice.edu/liquid-crystalline-order/.

Vinyl Institute. "PVC Materials in Building, Construction & Design." Accessed January 10, 2025. www.vinylinfo.org/uses/building-and-construction/.

Wyss Institute for Biologically Inspired Engineering at Harvard University. "Chitosan Bioplastic." Accessed January 10, 2025. https://wyss.harvard.edu/media-post/chitosan-bioplastic/.

Wyss Institute for Biologically Inspired Engineering at Harvard University. "Cry Me a River of Possibility: Scientists Design New Adaptive Material Inspired by Tears." Accessed January 10, 2025. https://wyss.harvard.edu/news/cry-me-a-river-of-possibility-scientists-design-new-adaptive-material-inspired-by-tears/.

Yeadon Space Agency. "Yeadon Space Agency Helps Storm Hoogland Create BioBench … Landscape Furniture That's Not Meant to Last." Accessed January 10, 2025. www.yeadonspaceagency.com/news/.

Yee, Maxine Swee-Li, Ling-Wei Hii, and Chin King Looi et al. "Impact of Microplastics and Nanoplastics on Human Health." *Nanomaterials (Basel, Switzerland)* 11, no. 2 (2021). https://doi.org/10.3390/nano11020496.

Zavada, Scott R., Nicholas R. McHardy, Keith L. Gordon, and Timothy F. Scott. "Rapid, Puncture-Initiated Healing via Oxygen-Mediated Polymerization." *ACS Macro Letters* 4, no. 8 (2015): 819–24. https://doi.org/10.1021/acsmacrolett.5b00315.

Zeo Sustainable Materials Pty Ltd. "Zeoform." Accessed January 10, 2025. https://zeoform.com/.

Zewe, Adam. "New 3D Printing Technique Creates Unique Objects Quickly and with Less Waste." Accessed January 10, 2025. https://news.mit.edu/2024/new-3d-printing-technique-creates-unique-objects-using-less-waste-1010.

4

SYNTOPIA 0—ANTHROPOS I HUMAN

From "Designer Matter" to "Critical Material"

Karola Dierichs

Syntopia—Notions

Syntopia 0—Anthropos I Human is a critical work reflecting on the surpassing of "all living biomass" by "anthropogenic mass" around the year 2020.[1] It is an architectural structure made largely from recycled plastic residue of the local Stuttgart car industry. Originally published as the *ICD Aggregate Pavilion 2018*, *Syntopia 0* has been re-named in retrospect (Figure 4.1). In this chapter, the genealogy of the project will be traced—from computational design to critical theory; from scientific images to artistic visions; from *dystopia* to *syntopia.*

The notion of "syntopy" has been introduced by the biologist Luis Rene Rivas in 1964.[2] He used the term to describe "two or more related species which occupy the same macrohabitat".[3] In an artistic-scientific context, the term "syntopia" has also been used to describe this interdisciplinary "fertilisation".[4]

The term "syntopy" is composed of the Greek words *syn**, meaning "together" and *topos*, meaning "place". It has—among others—been used to highlight sustainable projects specific to a region—a *topos*.[5] If applied to an architectural context, it denotes the process of an architectural structure being sourced from and embedded in its surrounding environment.[6] Here, the notion of "syntopy" is widened to encompass not only species co-habiting in one and the same place, but also material systems—such as architecture—entering into a symbiotic relationship with is ecosphere. The term also resonates with the notions of "utopia" and "dystopia". "Utopia", coined by Thomas More, describes a non-real place—"uo" in Greek denotes "non".[7] The word "dystopia" refers to a negative scenario.[8,9] Syntopia—on the contrary—is neither unreal nor negativistic. Its designs are based in reality taking it into preferable futures as outlined among others by Anthony Dunne and Fiona Raby.[10]

In this context, *Syntopia 0* is part of a series of architectural prototypes investigating the overall notion of "syntopy" by focussing on topical themes, such as the interrelation of body and architecture or the re-use of forest waste materials.[11,12] The central theme of *Syntopia 0*—the overturn of "all living biomass" by "anthropogenic mass"—is embedded in theories of the "Anthropocene", put forward among others by Paul Crutzen together with Eugene F. Stoer mer.[13,14,15,16] These theories suggest considering humanity as a major power impacting planet earth. Given *Syntopia 0* is made from recycled plastics from the Stuttgart car industry, it is an embodiment of human-produced waste.

DOI: 10.4324/9781003613589-5

FIGURE 4.1 The ICD Aggregate Pavilion 2018 has been developed within the paradigm of computational materials design. © Karola Dierichs, Institute for Comutational Design and Construction, University of Stuttgart.

The project will be traced from its origins in computational materials design towards its reflection in critical materials theory. In this context, the role of the image—in the scientific and artistic development stages of the project—will be discussed. As an outlook, future directions for *Syntopia* will be outlined.

Computational Materials Design

The notion of "designer matter" embraces the realms of physics and computational design. The term has been introduced by Pedro Reis, Heinrich M. Jaeger and Martin van Hecke.[17,18] This is the paradigm in which the project was perceived of as the *ICD Aggregate Pavilion 2018*.[19,20]

The architectural structure is made from a designed granular material.[21] Granular materials are defined as material systems consisting of large numbers of macroscopic particles between which mainly short-range contact forces are acting.[22,23,24,25,26] The sub-group of so-called designed granular materials belongs to a larger class of materials, often referred to as "designer matter".[27]

"Designer matter" describes materials in which the materiality and the form of the component elements at the mesoscale are defined to achieve specific functional effects on the macroscale of the material.[28,29] Related concepts are "smart matter", "active matter", or "metamatter", for example.[30] Transferring the notion of "designer matter" to granular materials, it follows that the individual particles in a granular material are designed to tune the behaviour of the overall material system.[31,32,33,34] Computational methods of optimization as well as fabrication are closely connected to the notion of materials design: on the one hand, "inverse" methods are deployed to find structural characteristics on a mesoscale to achieve a specific functionality on the macroscale[35]; on the other hand, rapid prototyping processes allow for fast production of materials with a wide spectrum of geometric variance.[36]

The *ICD Aggregate Pavilion 2018* was developed in three interrelated stages: (i) the design of the particles and the characterization of the granular materials made from them[37,38]; (ii) the investigation of spatial formations which can be made with these new materials[39,40]; (iii) the development and realization of a robotic construction system allowing for the particles to be poured on site and at full architectural scale.[41,42] The findings of these stages were integrated into one final demonstrator which was constructed in an industrial hall turned into a construction space.[43,44,45]

Based on preliminary experiments conducted in research and teaching, three main particle groups were identified: convex, non-convex and double non-convex.[46,47] For the realization of the project, initially only non-convex particle geometries were developed based on a parametric particle model. This model allowed for the definition of the "longest diameter of the convex hull, the arm amount, the axis length and the arm taper".[48] The particles were fabricated using selective laser sintering (SLS) for the initial tests with smaller particles and fewer amounts of particles (Figure 4.2). Departing from a norm particle, variations were formed changing arm amount and axis length. This allowed for studying how these variants affect the packing density and the compressive strength of the resultant granular materials.[49,50] One key finding was the fact that the variation of the arm length of a particle has a higher effect on the packing density of the granular material than a variation of its arm amount.[51] These findings allowed for defining the final geometry for an injection moulding process: a hexapod particle with 300 millimetres diameter of the convex hull and an axis length of 150 millimetres[52,53] (Figure 4.3a).

Spatial formations are based on the behaviour of the designed granular materials. These are mainly forms in which self-loads are transferred in compression, such as columns, arches and domes. These three structural types were the basis of the spatial investigations conducted both at the small model scale and at the large room scale. This allowed for establishing basic dimensions of these structures, such as the aspect ratio of columns or the span of an arch.[54] One key finding related to the formwork needed during the construction of an arch or dome: several options were investigated, such as clogging over a bottom hole, a large inflatable, bent wood sheets as well as smaller box units. Here, it became apparent that formwork made from smaller units leads to more reliably stable arches than formwork made from single large pieces.[55] Consequently, a formwork consisting of inflatable spheres—similar in size to that of the non-convex hexapod particle—was chosen: the geometrically convex spheres were light and could easily be removed, while the non-convex particles locked into place[56,57] (Figure 4.3b).

Granular materials are poured directly into the construction site. This is due to their capacity to re-configure based on external boundary conditions.[58,59,60,61] A suitable robotic system had to be developed alongside the designed granular material system—allowing for on-site and full-scale handling of the particles. For this, a cable-driven parallel robot was custom-made: four motors, placed on the ground, controlled an end-effector hanging on elevated cables.[62] The end-effector was a box which was also the storage unit of the particles—logistics and construction process were thus aligned[63] (Figure 4.4).

FIGURE 4.2 The packing density of granular materials consisting of designed particles with different extension lengths and amounts has been tested. © Karola Dierichs, Institute for Comutational Design and Construction, University of Stuttgart.

The *ICD Aggregate Pavilion 2018* was constructed in an industrial hall which was turned into a construction space by fitting the cable-driven parallel robot onto its walls and storing the boxes containing the particles directly next to the workspace of the robot.[64] The structure was designed based on two intersecting vaults—a "double-vault"[65]—which was formed by pouring hexapod particles in the zones where walls should be formed and spherical particles in those zones which should be empty space (Figure 4.5). The structure was built in layers with the cable-driven parallel robot run by a digital model. Once all layers had been laid, the spherical particles were removed from the inside of the structure and a double-vault remained standing.[66]

In this stage of the project as *ICD Aggregate Pavilion 2018*, it was perceived as an architectural research project in computational materials design in combination with large-scale construction robotics. It was therefore embedded in the realms of architecture and physics reflecting the concepts, methods and processes pertaining to these disciplines.

FIGURE 4.3 (a) Hexapods were made from recycled sprue from the local car industry. (b) Inflatable spheres were used as a removable formwork for a double-vaulted structure. © Karola Dierichs, Institute for Comutational Design and Construction, University of Stuttgart.

Critical Materials Theory

The finalization of the *ICD Aggregate Pavilion 2018* coincided with the exceeding of "all living biomass" by "anthropogenic mass" in the circa six years leading up to the year 2020.[67] This realization has led to a re-naming of the project in 2024: the *ICD Aggregate Pavilion 2018* became *Syntopia 0—Anthropos I Human.*[68]

This section will outline this becoming. It will introduce the notion of "critical materials" with respect to designed granular materials and its relation to *Syntopia* in architecture leading up the stage of the project as *Syntopia 0—Anthropos I Human.*

FIGURE 4.4 The structure was built using a cable-driven parallel robot which was suspended in a former industrial hall. © Karola Dierichs, Institute for Comutational Design and Construction, University of Stuttgart.

The particles used in the *ICD Aggregate Pavilion 2018* were made from so-called sprue. Sprue describes the channel through which plastic flows into a mould in an injection moulding process.[69] The material remaining in this channel is removed once the injection moulding process is finished and is thus a waste material.[70] For the designed particles in this project, sprue from a local manufacturer was used who works closely with the car industry in Stuttgart.[71]

In the reflection of the *ICD Aggregate Pavilion 2018* as part of the series on *Syntopia* in architecture, the recycled sprue material becomes "critical matter".[72] The notion of "critical matter" or "critical materialism" is closely linked to that of "new materialism": this idea considers matter an essential actor in the construction of our realities; it encompasses the notions of matter as being active, of biological matter and of its socioeconomic implication.[73] Often it also refers to rare materials and their logistics.[74] A seminal account of this latter category is given in Anna L. Tsing's *The Mushroom at the End of the World.*[75] Several research groups are starting to develop their agenda around "critical materials".[76] "Critical matter" is part of a larger movement calling for a field of "critical post-humanities".[77] One core paradigm is that "the Critical Posthumanities no longer assume that the subject of knowledge is *homo universalis* or Anthropos".[78] They counter this with a "transversal knowledge-production entity" including "human and nonhuman" entities.[79] The notion of *Syntopia* is instilled with this discourse—it is an

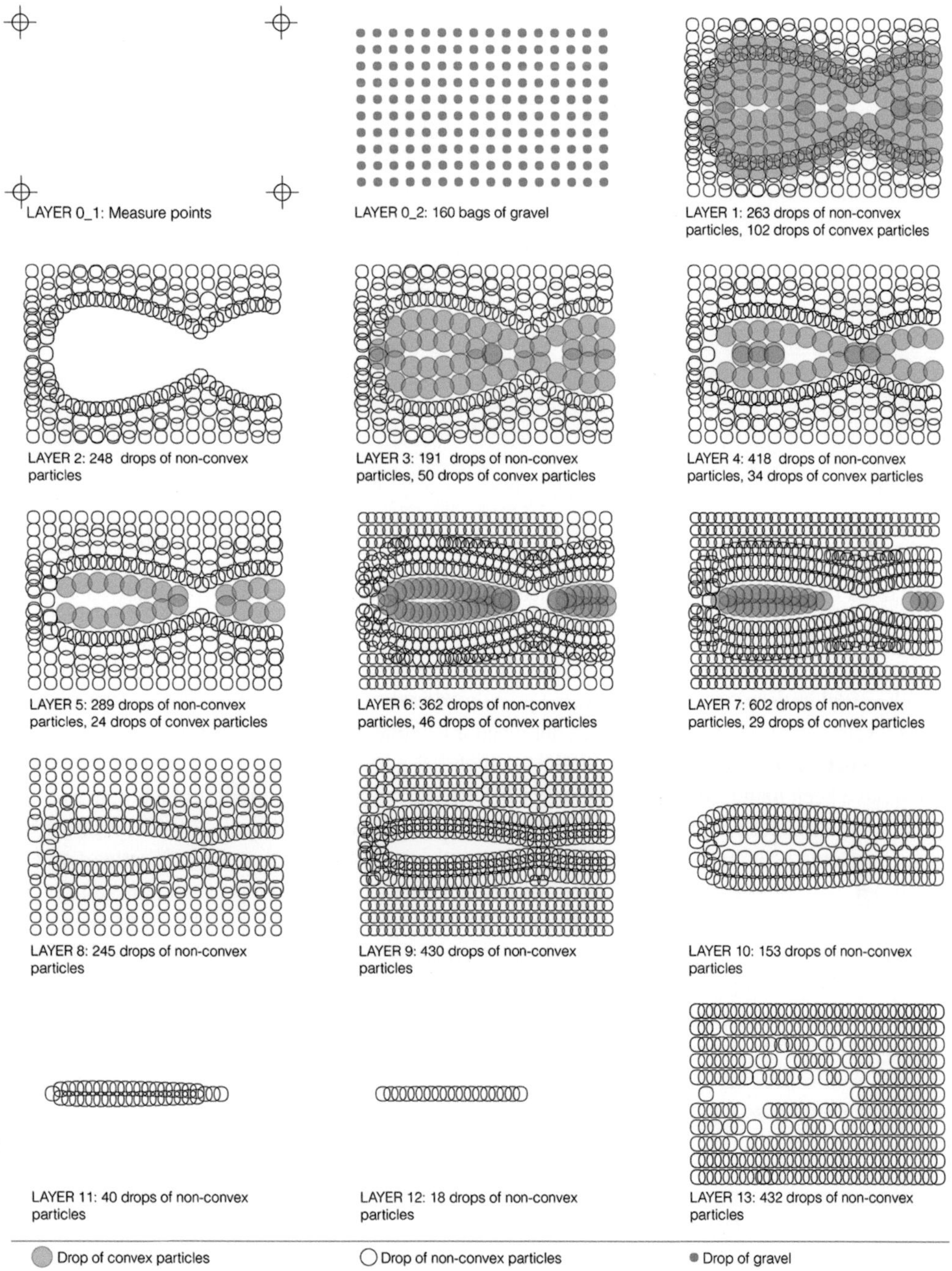

FIGURE 4.5 All construction steps have been recorded digitally including the location and number of drops for non-convex and convex particles. © Karola Dierichs, Institute for Comutational Design and Construction, University of Stuttgart.

Source: Karola Dierichs and Achim Menges, "Emergent Enclosures: Granular Architectures," in *Digital Fabrication in Interior Design: Body, Object, Enclosure*, ed. Jonathon Anderson and Lois Weinthal (Routledge, 2021).

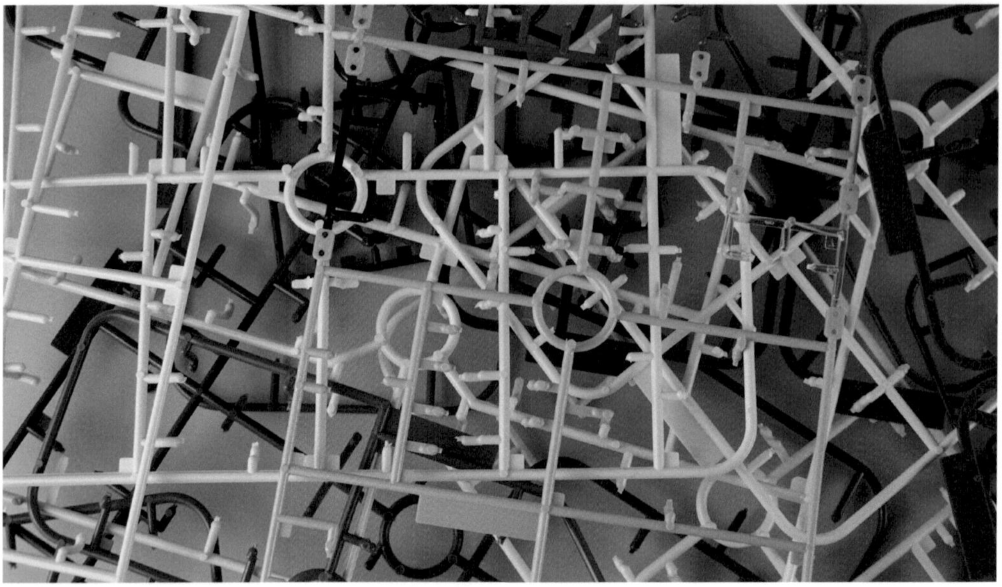

FIGURE 4.6 Leftover from injection moulded pieces including sprue are usually considered a waste byproduct. © Rian Arif I Shutterstock.

architectural translation of the multiverse postulated in the current-day humanities, interrelating the sourcing and processing of matter, the place it is situated in and its inhabitants

Considering the plastic sprue material—which the particles of the *ICD Aggregate Pavilion 2018* have been made from—in the context of "critical matter", one might find that it is a direct product of the "Plasticene Epoch" which the New York Times called out in 2014[80] (Figure 4.6). It logs into the economic value chain of the car industry, local to the region of Stuttgart and is a register of its production processes contributing to plastic waste. The project is thus aiming at critically assessing and revaluing this waste material with a focus on its underlying economic processes.[81] It therefore reflects on the close interrelation between "Anthropocene" and the idea of a "Capitalocene".[82]

Naming the project *Syntopia 0—Anthropos I Human* situates the former *ICD Aggregate Pavilion 2018* in the context of the discourse on "critical matter". It acknowledges that "syntopy" in the "Anthropocene" denotes working and living with human-produced waste products—embracing and possibly transforming the uncanniness of a *topos* that is filled with our own debris.

Scientific Images—Artistic Visions

In both its stages—that of the *ICD Aggregate Pavilion 2018* and that of *Syntopia 0—Anthropos I Human*—visual representations of the work played a key role.

In the first stage, photographic images of physical experiments alongside data collection were key to observing and analysing the properties and behaviour of the designed granular materials. These are complemented by and compared to numerical simulations in a Discrete Element Method (DEM) software package[83] (Figure 4.7). A special form of analytical image-generation

FIGURE 4.7 Particle designs have been analysed using experiments and simulations. © Karola Dierichs with ITASCA Education Partnership (IEP) Research Program, Institute for Comutational Design and Construction, University of Stuttgart.

were photo-elastic experiments showing force-lines in a two-dimensional designed granular material—these were conducted at the Behringer Lab of Duke University.[84]

These scientific images have the epistemic potential of cognition—of learning about the behaviour of a designed granular material based on hypotheses which are mostly established a priori. Yet this process is non-linear: especially in early stages, physical experiments and digital simulations render surprising results requiring to move back and forth between evaluations and hypotheses. In all cases, the images taken from the physical experiments were given a higher epistemic value than those stemming from numerical simulations. This is due to the fact that granular materials have a high degree of statistical variance in their behaviour so that small causes can accumulate to larger effects—simulations are reductionist and might leave out some of these small causes which in the end are critical: in his essay "The Computational Fallacy", Sanford Kwinter states that "no computer on earth can match of even the most simple natural system, (...)."[85] Eric Winsberg poses a slightly different, yet complementary view on the interrelation of experiment and simulation in his essay "A Tale of Two Methods".[86] While he acknowledges that experiments are "epistemologically prior to simulations" they are in his view not "more epistemically powerful".[87] The key in both arguments is the presence of an "intrinsic" understanding or "sufficient systematic knowledge" of a given physical phenomenon.[88]

Considering this, the process of establishing a characterization of the materials started from physical experimentation which was then selectively compared to numerical simulations. These latter ones were used to gain additional information which in a physical experiment is hard or impossible to gather such as the number of contact points between particles.[89]

After completion, the project was photographed by Roland Halbe.[90] In this artistic collaboration images are not mere representatives of a scientific outcome but rather incorporations of underlying design visions—these emerge during the making of the images—often not pre-known image stories emerge between the creators, the photographed architectural object, the human models and the photographer. This process closely relates to the notion of the "Bildakt"—the inner activity of images.[91] Here, images are no longer seen as passive forms of representations but rather as active epistemic and hermeneutic entities.[92] This entails the notion of "mnemosyne"—which is the Greek name of the goddess of memory, as well as the title of Aby Warburg's topical image collection: images are considered to be stored in our individual

FIGURE 4.8 In 2024,the ICD Aggregate Pavilion 2018 has been renamed *Syntopia 0–Anthropos I Human* following the overturn of "all living biomass" by "anthropogenic mass" (Elhacham et al. 2020). © Roland Halbe, Roland Halbe Architectural Photography.

and collective memory similar to words and thus re-emerge time and again in many variations of a theme.[93]

One of the photographs in particular envisions the becoming of *Syntopia 0—Anthropos I Human* (Figure 4.8). It is a centrally symmetric interior view of the architectural structure made from the hexapodal designed particles. Daylight enters the space from top where the granular vault becomes thinner—giving the scene a cathedral-like atmosphere. The ground is covered in white-greyish granules—marble. Two figures stand in this room, their back turned towards the spectator. One of them is dressed in light grey, the other in black—the colours alluding to opposing pairs: the light and the dark—the angel and the devil—the good and the bad—life and death. Yet these are no longer separate but unified in a common gaze at the structure made from human debris which is at once threatening and enlightening—the white particles appearing as stars from the "plasticene epoch".[94] Here, the image is an active agent in forming an artistic vision of the project transforming from a merely scientific investigation into *Syntopia 0—Anthropos I Human.*

Syntopia—Futures

While *Syntopia* is a concept applying to many *topoi*—many places—its relevance to the age of plastics and its relating counterpart the "Anthropocene" is ambivalent. The syllable *syn** evokes a harmonious co-creation, alluding to the thought that an architectural structure is made from its surrounding nature—and one might most immediately perceive of that nature as biological. A famous depiction is that of the Vitruvian Hut by Charles Eisen.[95] But what if plastic was already an intrinsic part of this very nature that we embed ourselves in? A current-day notion of *Syntopia* needs to include the post-human materials landscape into its designs—at the least for critical discussion, at the best for their re-valuation.

Acknowledgments

Karola Dierichs acknowledges the support of the Cluster of Excellence *Matters of Activity. Image Space Material* funded by the Deutsche Forschungsgemeinschaft (DFG, German Research Foundation) under Germany's Excellence Strategy—EXC 2025—390648296.

Project Credits

SYNTOPIA 0—ANTHROPOS I HUMAN
Karola Dierichs, Matters of Activity, Humboldt-Universität zu Berlin

FORMERLY KNOWN AS ICD AGGREGATE PAVILION 2018

ICD Institute for Computational Design and Construction
Karola Dierichs, Achim Menges

Research Assistants
Christian Arias, Bahar Al Bahar, Elaine Bonavia, Federico Forestiero, Pedro Giachini, Shir Katz, Alexandre Mballa- Ekobena, Leyla Yunis, Jacob Zindroski

Cable Robotics
Ondřej Kyjánek, Martin Loučka

Manufacturing
Wilhelm Weber GmbH & Co. KG

Funding
GETTYLAB
ITASCA Consulting Inc.

SYNTOPIA 0 HAS BEEN PART OF BEING PLASTIC/BECOMING PLASTIC

Team Director
Assistant Prof. Dr.- Ing. Ehsan Baharlou

Research Assistants
Avery Edson, Keaton Fisher (former researcher), Eli Sobel, Tabi Summers

Notes

1 Emily Elhacham et al., "Global Human-Made Mass Exceeds All Living Biomass," *Nature* 588, no. 7838 (2020).

2 Luis R. Rivas, "A Reinterpretation of the Concepts "Sympatric" and "Allopatric" with Proposal of the Additional Terms "Syntopic" and "Allotopic"," *Systematic Biology* 13, 1-4 (1964), https://doi.org/10.2307/sysbio/13.1-4.42.

3 Rivas, "A Reinterpretation of the Concepts "Sympatric" and "Allopatric" with Proposal of the Additional Terms "Syntopic" and "Allotopic"," 43.

4 Andrea Niehaus and Dieter Ronte, eds, *Syntopia* (Wienand, 2007) Igor Sacharow-Ross.

5 Boris Previsic, ed., *Syntopia Alpina-online Magazine* (2025), www.syntopia-alpina.ch/, accessed February 17, 2025.

6 "Syntopic Architectures," Matters of Activity, accessed February 17, 2025, www.matters-of-activity.de/en/posts/12192/syntopic-architectures.

7 Simon Blackburn, *Utopia* (Oxford University Press, 2016), accessed June 15, 2025, www.oxfordreference.com/view/10.1093/acref/9780198735304.001.0001/acref-9780198735304-e-3216.

8 Dinah Birch and Katy Hooper, *Dystopia* (Oxford University Press, 2013), accessed June 15, 2025, www.oxfordreference.com/view/10.1093/acref/9780199608218.001.0001/acref-9780199608218-e-2404.

9 George M. Logan et al., eds, *Thomas More: Utopia* (Cambridge University Press, 2008).

10 Anthony Dunne and Fiona Raby, *Speculative Everything: Design, Fiction and Social Dreaming* (The MIT Press, 2013), 5.

11 Elaine Bonavia et al., "Minimal Machines: Augmented Reality for Filament-Construction of Partially Ordered Systems in Architecture," *Construction Robotics* 7, no. 3 (2023).

12 Pelin Asa et al., "Architectures of Syntopia: An Interdisciplinary Speculative Model for Constructions with Insect-Infested Wood," *TATuP – Zeitschrift für Technikfolgenabschätzung in Theorie und Praxis* 33, no. 3 (2024), https://doi.org/10.14512/tatup.7153.

13 Elhacham et al., "Global human-made mass exceeds all living biomass."

14 Paul J. Crutzen and Eugene F. Stoermer, "The 'Anthropocene'," *The International Geosphere–Biosphere Programme (IGBP) Newsletter* 41 (2000).

15 Paul J. Crutzen, "Geology of Mankind," *Nature* 415, no. 6867 (2002), https://doi.org/10.1038/415023a.

16 Paul J. Crutzen and Eugene F. Stoermer, "The 'Anthropocene' (2000)," in *The Future of Nature: Documents of Global Change*, ed. Libby Robin, Sverker Sörlin and Paul Warde (Yale University Press, 2013).

17 Pedro M. Reis et al., "Designer Matter: A Perspective," *Extreme Mechanics Letters* 5 (2015), https://doi.org/10.1016/j.eml.2015.09.004,www.sciencedirect.com/science/article/pii/S2352431615300080.

18 Heinrich M. Jaeger, "Celebrating Soft Matter's 10th Anniversary: Toward Jamming by Design," *Soft Matter* 11, no. 1 (2015), https://doi.org/10.1039/C4SM01923G.

19 Karola Dierichs and Achim Menges, "Designing Architectural Materials: From Granular Form to Functional Granular Material," *Bioinspiration & Biomimetics* 16, no. 6 (2021), https://doi.org/10.1088/1748-3190/ac2987.

20 Karola Dierichs and Achim Menges, "Towards an Aggregate Architecture: Designed Granular Systems as Programmable Matter in Architecture," *Granular Matter* 18, no. 2 (2016), https://doi.org/10.1007/s10035-016-0631-3.

21 Dierichs and Menges, "Designing architectural materials: from granular form to functional granular material."

22 Jacques Duran, *Sands, Powders, and Grains: An Introduction to the Physics of Granular Materials (1 Ed.),* Partially Ordered Systems (Springer-Verlag, 2000).

23 Pierre-Gilles de Gennes, "Granular Matter: A Tentative View," *Rev. Mod. Phys.* 71, no. 2 (1999), https://doi.org/10.1103/RevModPhys.71.S374, http://link.aps.org/doi/10.1103/RevModPhys.71.S374.

24 Pierre-Gilles de Gennes, "Reflections on the Mechanics of Granular Matter," *Physica A: Statistical Mechanics and its Applications* 261, no. 3 (1998), https://doi.org/10.1016/S0378-4371(98)00438-5,www.sciencedirect.com/science/article/pii/S0378437198004385.

25 Heinrich M. Jaeger et al., "Granular Solids, Liquids, and Gases," *Rev. Mod. Phys.* 68, no. 4 (1996), https://doi.org/10.1103/RevModPhys.68.1259,https://link.aps.org/doi/10.1103/RevModPhys.68.1259.

26 Heinrich M. Jaeger et al., "The Physics of Granular Materials," *Physics Today* 49, no. 4 (1996), https://doi.org/10.1063/1.881494.

27 Reis et al., "Designer Matter: A perspective."
28 Reis et al., "Designer Matter: A perspective."
29 Jaeger, "Celebrating Soft Matter's 10th Anniversary: Toward jamming by design."
30 Reis et al., "Designer Matter: A perspective," 25.
31 Dierichs and Menges, "Towards an aggregate architecture: designed granular systems as programmable matter in architecture."
32 Jaeger, "Celebrating Soft Matter's 10th Anniversary: Toward jamming by design."
33 Michael Hensel and Achim Menges, "Material Systems - Introduction: 20 Proto-Architectures, Research and Design Projects," in Hensel; Menges, *Morpho-Ecologies*.
34 Michael Hensel and Achim Menges, "Eiichi Matsuda - Aggregates 01 - 2003-2004 [Project Description]," in Hensel; Menges, *Morpho-Ecologies*.
35 Jaeger, "Celebrating Soft Matter's 10th Anniversary: Toward jamming by design."
36 Reis et al., "Designer Matter: A perspective."
37 Dierichs and Menges, "Designing architectural materials: from granular form to functional granular material."
38 Karola Dierichs, *Granular Architectures-Granular Materials as »Designer Matter« in Architecture* (Institute for Computational Design and Construction, University of Stuttgart, 2020), 168–171, 192–204, and 256–267.
39 Dierichs and Menges, "Designing architectural materials: from granular form to functional granular material."
40 Dierichs, Granular Architectures-Granular Materials as »Designer Matter« in Architecture, 211–25.
41 Dierichs, *Granular Architectures-Granular Materials as »Designer Matter« in Architecture*, 179–180 and 226.
42 Karola Dierichs et al., "Construction Robotics for Designed Granular Materials: In Situ Construction with Designed Granular Materials at Full Architectural Scale Using a Cable-Driven Parallel Robot," *Construction Robotics* 3, no. 1 (2019), https://doi.org/10.1007/s41693-019-00024-6.
43 Dierichs and Menges, "Designing architectural materials: from granular form to functional granular material."
44 Dierichs, *Granular Architectures-Granular Materials as »Designer Matter« in Architecture*, 232–49.
45 Dierichs et al., "Construction robotics for designed granular materials: in situ construction with designed granular materials at full architectural scale using a cable-driven parallel robot."
46 Dierichs and Menges, "Designing architectural materials: from granular form to functional granular material," 9–11.
47 Dierichs, *Granular Architectures-Granular Materials as »Designer Matter« in Architecture*, 138–41.
48 Dierichs, *Granular Architectures-Granular Materials as »Designer Matter« in Architecture*, 168–71.
49 Dierichs and Menges, "Designing architectural materials: from granular form to functional granular material," 8.
50 Dierichs, *Granular Architectures-Granular Materials as »Designer Matter« in Architecture*, 192–204.
51 Dierichs, *Granular Architectures-Granular Materials as »Designer Matter« in Architecture*, 200.
52 Dierichs and Menges, "Designing architectural materials: from granular form to functional granular material," 13–14.
53 Dierichs, *Granular Architectures-Granular Materials as »Designer Matter« in Architecture*, 173–75.
54 Dierichs, *Granular Architectures-Granular Materials as »Designer Matter« in Architecture*, 175–178 and 211–225.
55 Dierichs, *Granular Architectures-Granular Materials as »Designer Matter« in Architecture*, 213 Figure 9.20.
56 Dierichs, *Granular Architectures-Granular Materials as »Designer Matter« in Architecture*, 256–61.
57 Dierichs et al., "Construction robotics for designed granular materials: in situ construction with designed granular materials at full architectural scale using a cable-driven parallel robot," 44–45.
58 Dierichs, *Granular Architectures-Granular Materials as »Designer Matter« in Architecture*, 10.
59 Sean Keller and Heinrich M. Jaeger, "Aleatory Architectures," *Granular Matter* 18, no. 2 (2016): 29.3, https://doi.org/10.1007/s10035-016-0629-x.
60 Karola Dierichs and Achim Menges, "Aggregate Structures: Material and Machine Computation of Designed Granular Substances," *Architectural Design* 82, no. 2 (2012): 76–77, https://doi.org/10.1002/ad.1382.
61 Michael Hensel and Achim Menges, "Aggregates," *Architectural Design* 78, no. 2 (2008): 82, https://doi.org/10.1002/ad.645.

62 Dierichs, *Granular Architectures-Granular Materials as »Designer Matter« in Architecture*, 179–180 and 226.
Dierichs et al., "Construction robotics for designed granular materials: in situ construction with designed granular materials at full architectural scale using a cable-driven parallel robot."
63 Dierichs, *Granular Architectures-Granular Materials as »Designer Matter« in Architecture*, 227–31.
Dierichs et al., "Construction robotics for designed granular materials: in situ construction with designed granular materials at full architectural scale using a cable-driven parallel robot."
64 Dierichs, *Granular Architectures-Granular Materials as »Designer Matter« in Architecture*, 232–34.
Dierichs et al., "Construction robotics for designed granular materials: in situ construction with designed granular materials at full architectural scale using a cable-driven parallel robot."
65 Dierichs, *Granular Architectures-Granular Materials as »Designer Matter« in Architecture*, 234–39.
66 Dierichs, *Granular Architectures-Granular Materials as »Designer Matter« in Architecture*, 239–48.
67 Elhacham et al., "Global human-made mass exceeds all living biomass."
68 "Syntopia 0-Anthropos I Human," Matters of Activity, accessed March 30, 2025, www.matters-of-activity.de/en/activities/13470/syntopia-0-anthropos-i-human.
69 Marcel Escudier and Tony Atkins, "Sprue," in *A Dictionary of Mechanical Engineering* (Oxford University Press, 2019), accessed March 23, 2025, www.oxfordreference.com/view/10.1093/acref/9780198832102.001.0001/acref-9780198832102-e-6038.
70 "Sprue," Merriam-Webster, accessed June 15, 2025, www.merriam-webster.com/dictionary/sprue.
71 Weber, "Weber Mehrkomponententechnik," 2025, accessed June 15, 2025, www.weber-esslingen.de/de/.
72 Rosi Braidotti, *Posthuman Knowledge and the Critical Posthumanities,* The Incidents (Harvard Design Press; Sternberg Press, 2024), 21–24.
Tobias Goll et al., eds, *Critical Matter: Diskussionen Eines Neuen Materialismus* (Edition Assemblage, 2014).
Diana Coole and Samantha Frost, "Introducing the New Materialisms," in *New Materialisms: Ontology, Agency, and Politics*, ed. Diana Coole and Samantha Frost (Duke University Press, 2010), 24–36.
73 Coole and Frost, "Introducing the new materialisms."
74 Lamia Mouloudi and Karine Evrard Samuel, "Critical Materials: A Systematic Literature Review," *SSRN*, 2022, https://doi.org/10.2139/ssrn.4108632.
75 Anna Lowenhaupt Tsing, *The Mushroom at the End of the World: On the Possibility of Life in Capitalist Ruins* (Princeton University Press, 2015).
76 "Critical Matter: Designing Across the Scale, Imagining a Transformative Future," MIT Media Lab, accessed March 29, 2025, www.media.mit.edu/groups/critical-matter/overview/.
"SustainLab: Platform for Critical Matter," Burg Giebichenstein Kunsthochschule Halle, accessed March 29, 2025, www.youtube.com/watch?v=uuQMwdUs2Dg.
77 Braidotti, *Posthuman knowledge and the critical posthumanities*, 43–50.
78 Braidotti, *Posthuman knowledge and the critical posthumanities*, 48.
79 Braidotti, *Posthuman knowledge and the critical posthumanities*, 49.
80 "Notes from the Plasticene Epoch," New York Times Editorial Board, accessed March 23, 2025, www.nytimes.com/2014/06/15/opinion/sunday/from-ocean-to-beach-tons-of-plastic-pollution.html.
81 Goll et al., *Critical Matter: Diskussionen eines neuen Materialismus.*
Coole and Frost, "Introducing the new materialisms," 24–36.
82 Jason W. Moore, ed., *Anthropocene or Capitalocene? Nature, History, and the Crisis of Capitalism* (Kairos Books, PM Press, 2016).
83 Dierichs and Menges, "Towards an aggregate architecture: designed granular systems as programmable matter in architecture," 4.
Dierichs and Menges, "Designing architectural materials: from granular form to functional granular material," 7–8.
84 Karola Dierichs and Achim Menges, "Granular Morphologies: Programming Material Behaviour with Designed Aggregates," *Architectural Design* 85, no. 5 (2015): 88, https://doi.org/10.1002/ad.1959.
85 Sanford Kwinter, "The Computational Fallacy," *Thresholds*, no. 26 (2003): 91, accessed April 1, 2025, www.jstor.org/stable/43876166.
86 Eric Winsberg, "A Tale of Two Methods," *Synthese* 169, no. 3 (2009).
87 Winsberg, "A tale of two methods," 591.
88 Kwinter, "The Computational Fallacy," 91.
Winsberg, "A tale of two methods," 591.

89 Dierichs and Menges, "Designing architectural materials: from granular form to functional granular material," 8.
90 "ICD Aggregate Pavilion 2018," Roland Halbe Architectural Photography, accessed March 30, 2025, https://rolandhalbe.eu/portfolio/aggregate-pavilion-by-icd-university-of-stuttgart/.
91 Horst Bredekamp, *Der Bildakt* (Verlag Klaus Wagenbach, 2015).
92 Bredekamp, *Der Bildakt*, 60.
93 "Aby Warburg Bilderatlas Mnemosyne Virtual Exhibition," The Warburg Institute, accessed June 15, 2025, https://warburg.sas.ac.uk/aby-warburg-bilderatlas-mnemosyne-virtual-exhibition.
94 New York Times Editorial Board, "Notes From the Plasticene Epoch."
95 Marc Antoine Laugier, *Essai Sur L'architecture*, 2nd ed. (Paris: Chez Duchesne, 1755), accessed April 1, 2025. https://doi.org/10.3931/e-rara-128.

References

Asa, Pelin, Karin Krauthausen, Robert Stock, and Karola Dierichs. "Architectures of Syntopia: An Interdisciplinary Speculative Model for Constructions with Insect-Infested Wood." *TATuP – Zeitschrift für Technikfolgenabschätzung in Theorie und Praxis* 33, no. 3 (2024): 48–56. https://doi.org/10.14512/tatup.7153.

Birch, Dinah, and Katy Hooper. *Dystopia*. Oxford University Press, 2013. Accessed June 15, 2025. www.oxfordreference.com/view/10.1093/acref/9780199608218.001.0001/acref-9780199608218-e-2404.

Bonavia, Elaine, Jessica Farmer, Alexandre Mballa-Ekobena, Nikolai Rosenthal, Laurence Douny, and Karola Dierichs. "Minimal Machines: Augmented Reality for Filament-Construction of Partially Ordered Systems in Architecture." *Construction Robotics* 7, no. 3 (2023): 329–50.

Braidotti, Rosi. *Posthuman Knowledge and the Critical Posthumanities*. The Incidents. Harvard Design Press; Sternberg Press, 2024.

Bredekamp, Horst. *Der Bildakt*. Verlag Klaus Wagenbach, 2015.

Burg Giebichenstein Kunsthochschule Halle. "SustainLab: Platform for Critical Matter." Accessed March 29, 2025. www.youtube.com/watch?v=uuQMwdUs2Dg.

Coole, Diana, and Samantha Frost. "Introducing the New Materialisms." In *New Materialisms: Ontology, Agency, and Politics*. Edited by Diana Coole and Samantha Frost, 1–43. Duke University Press, 2010.

Crutzen, Paul J. "Geology of Mankind." *Nature* 415, no. 6867 (2002): 23. https://doi.org/10.1038/415023a.

Crutzen, Paul J., and Eugene F. Stoermer. "The 'Anthropocene'." *The International Geosphere–Biosphere Programme (IGBP) Newsletter* 41 (2000): 17–18.

Crutzen, Paul J., and Eugene F. Stoermer. "The 'Anthropocene' (2000)." In *The Future of Nature: Documents of Global Change*. Edited by Libby Robin, Sverker Sörlin and Paul Warde, 479–90. Yale University Press, 2013.

de Gennes, Pierre-Gilles. "Reflections on the Mechanics of Granular Matter." *Physica A: Statistical Mechanics and its Applications* 261, no. 3 (1998): 267–93. https://doi.org/10.1016/S0378-4371(98)00438-5. www.sciencedirect.com/science/article/pii/S0378437198004385.

de Gennes, Pierre-Gilles. "Granular Matter: A Tentative View." *Reviews of Modern Physics* 71, no. 2 (1999): S374–S382. https://doi.org/10.1103/RevModPhys.71.S374. http://link.aps.org/doi/10.1103/RevModPhys.71.S374.

Dierichs, Karola. *Granular Architectures-Granular Materials as »Designer Matter« in Architecture*. Institute for Computational Design and Construction, University of Stuttgart, 2020.

Dierichs, Karola, and Achim Menges. "Aggregate Structures: Material and Machine Computation of Designed Granular Substances." *Architectural Design* 82, no. 2 (2012): 74–81. https://doi.org/10.1002/ad.1382.

Dierichs, Karola, and Achim Menges. "Granular Morphologies: Programming Material Behaviour with Designed Aggregates." *Architectural Design* 85, no. 5 (2015): 86–91. https://doi.org/10.1002/ad.1959.

Dierichs, Karola, and Achim Menges. "Towards an Aggregate Architecture: Designed Granular Systems as Programmable Matter in Architecture." *Granular Matter* 18, no. 2 (2016): 25. https://doi.org/10.1007/s10035-016-0631-3.

Dierichs, Karola, and Achim Menges. "Designing Architectural Materials: From Granular Form to Functional Granular Material." *Bioinspiration & Biomimetics* 16, no. 6 (2021): 65010. https://doi.org/10.1088/1748-3190/ac2987.

Dierichs, Karola, and Achim Menges. "Emergent Enclosures: Granular Architectures." In *Digital Fabrication in Interior Design: Body, Object, Enclosure*. Edited by Jonathon Anderson and Lois Weinthal, 157–74. Routledge, 2021.

Dierichs, Karola, Ondřej Kyjánek, Martin Loučka, and Achim Menges. "Construction Robotics for Designed Granular Materials: In Situ Construction with Designed Granular Materials at Full Architectural Scale Using a Cable-Driven Parallel Robot." *Construction Robotics* 3, no. 1 (2019): 41–52. https://doi.org/10.1007/s41693-019-00024-6.

Dunne, Anthony, and Fiona Raby. *Speculative Everything: Design, Fiction and Social Dreaming*. The MIT Press, 2013.

Duran, Jacques. *Sands, Powders, and Grains: An Introduction to the Physics of Granular Materials (1 Ed.)*. Partially Ordered Systems. Springer-Verlag, 2000.

Elhacham, Emily, Liad Ben-Uri, Jonathan Grozovski, Yinon M. Bar-On, and Ron Milo. "Global Human-Made Mass Exceeds All Living Biomass." *Nature* 588, no. 7838 (2020): 442–44.

Escudier, Marcel, and Tony Atkins. "Sprue." In *A Dictionary of Mechanical Engineering*. Oxford University Press, 2019. Accessed March 23, 2025. www.oxfordreference.com/view/10.1093/acref/9780198832102.001.0001/acwref-9780198832102-e-6038.

Goll, Tobias, Daniel Keil, and Thomas Telios, eds. *Critical Matter: Diskussionen Eines Neuen Materialismus*. Edition Assemblage, 2014.

Hensel, Michael, and Achim Menges. "Eiichi Matsuda - Aggregates 01 - 2003-2004 [Project Description]." In *Morpho-Ecologies*. Edited by Michael Hensel and Achim Menges, 262–71. Architectural Association Publications, 2006.

Hensel, Michael, and Achim Menges. "Material Systems - Introduction: 20 Proto-Architectures, Research and Design Projects." In *Morpho-Ecologies*. Edited by Michael Hensel and Achim Menges, 62–67. Architectural Association Publications, 2006.

Hensel, Michael, and Achim Menges, eds. *Morpho-Ecologies*. AA Publications, 2006.

Hensel, Michael, and Achim Menges. "Aggregates." *Architectural Design* 78, no. 2 (2008): 80–87. https://doi.org/10.1002/ad.645.

Jaeger, Heinrich M. "Celebrating Soft Matter's 10th Anniversary: Toward Jamming by Design." *Soft Matter* 11, no. 1 (2015): 12–27. https://doi.org/10.1039/C4SM01923G.

Jaeger, Heinrich M., Sidney R. Nagel, and Robert P. Behringer. "Granular Solids, Liquids, and Gases." *Reviews of Modern Physics* 68, no. 4 (1996): 1259–73. https://doi.org/10.1103/RevModPhys.68.1259. https://link.aps.org/doi/10.1103/RevModPhys.68.1259.

Jaeger, Heinrich M., Sidney R. Nagel, and Robert P. Behringer. "The Physics of Granular Materials." *Physics Today* 49, no. 4 (1996): 32–38. https://doi.org/10.1063/1.881494.

Keller, Sean, and Heinrich M. Jaeger. "Aleatory Architectures." *Granular Matter* 18, no. 2 (2016): 29. https://doi.org/10.1007/s10035-016-0629-x.

Kwinter, Sanford. "The Computational Fallacy." *Thresholds* 26 (2003): 90–92. Accessed April 1, 2025. www.jstor.org/stable/43876166.

Laugier, Marc Antoine. *Essai Sur L'architecture*. 2nd ed. Chez Duchesne, 1755. Accessed April 1, 2025. https://doi.org/10.3931/e-rara-128.

Logan, George M., Robert M. Adams, and Clarence H. Miller, eds. *Thomas More: Utopia*. Cambridge University Press, 2008.

Matters of Activity. "Syntopic Architectures." Accessed February 17, 2025. www.matters-of-activity.de/en/posts/12192/syntopic-architectures.

Matters of Activity. "Syntopia 0-Anthropos I Human." Accessed March 30, 2025. www.matters-of-activity.de/en/activities/13470/syntopia-0-anthropos-i-human.

Merriam-Webster. "Sprue." Accessed June 15, 2025. www.merriam-webster.com/dictionary/sprue.

MIT Media Lab. "Critical Matter: Designing Across the Scale, Imagining a Transformative Future." Accessed March 29, 2025. www.media.mit.edu/groups/critical-matter/overview/.

Moore, Jason W., ed. *Anthropocene or Capitalocene? Nature, History, and the Crisis of Capitalism*. Kairos Books, PM Press, 2016.

Mouloudi, Lamia, and Karine Evrard Samuel. "Critical Materials: A Systematic Literature Review." *SSRN*, 2022. https://doi.org/10.2139/ssrn.4108632.

New York Times Editorial Board. "Notes from the Plasticene Epoch." Accessed March 23, 2025. www.nytimes.com/2014/06/15/opinion/sunday/from-ocean-to-beach-tons-of-plastic-pollution.html.

Niehaus, Andrea, and Dieter Ronte, eds. *Syntopia*. With the assistance of Igor Sacharow-Ross. Wienand, 2007.

Previsic, Boris, ed. *Syntopia Alpina-online Magazine*. 2025. Accessed February 17, 2025. www.syntopia-alpina.ch/.

Reis, Pedro M., Heinrich M. Jaeger, and Martin van Hecke. "Designer Matter: A Perspective." *Extreme Mechanics Letters* 5 (2015): 25–29. https://doi.org/10.1016/j.eml.2015.09.004. www.sciencedirect.com/science/article/pii/S2352431615300080.

Rivas, Luis Rene. "A Reinterpretation of the Concepts "Sympatric" and "Allopatric" with Proposal of the Additional Terms "Syntopic" and "Allotopic."" *Systematic Biology* 13, no. 1-4 (1964): 42–43. https://doi.org/10.2307/sysbio/13.1-4.42.

Roland Halbe Architectural Photography. "ICD Aggregate Pavilion 2018." Accessed March 30, 2025. https://rolandhalbe.eu/portfolio/aggregate-pavilion-by-icd-university-of-stuttgart/.

Simon Blackburn. *Utopia*. Oxford University Press, 2016. Accessed June 15, 2025. www.oxfordreference.com/view/10.1093/acref/9780198735304.001.0001/acref-9780198735304-e-3216.

The Warburg Institute. "Aby Warburg Bilderatlas Mnemosyne Virtual Exhibition." Accessed June 15, 2025. https://warburg.sas.ac.uk/aby-warburg-bilderatlas-mnemosyne-virtual-exhibition.

Tsing, Anna Lowenhaupt. *The Mushroom at the End of the World: On the Possibility of Life in Capitalist Ruins*. Princeton University Press, 2015.

Weber. "Weber Mehrkomponententechnik." 2025. Accessed June 15, 2025. www.weber-esslingen.de/de/.

Winsberg, Eric. "A Tale of Two Methods." *Synthese* 169, no. 3 (2009): 575–92.

5

BREATHING ENCLOSURES AND ENGINEERED LIVING MATERIALS

Maria Paz Gutierrez

Introduction

Incorporating material breathability as a climatic control strategy is as old as construction. Primitive tents from animal—and plant-derived membranes attest to these early efforts.[1] Whether in the form of tensile fabrics or air-inflated skins, architecture has used elastic properties as a primary means to explore the regulations of structures, temporality, and climate. From early on, cultures have empirically explored opportunities of using air from supporting rituals to survival strategies for buoyancy. Elastic materials and fabrication techniques continue to fuel innovative concepts and developments in inhabitation and enclosure design. From developing phased elasticity to 3D printing materials with selective flexibility, architects seek to establish new architectural opportunities with elastomeric polymers, and more recently, biopolymers.[2]

Various materials, functions and forms have historically incorporated flexible membranes in construction (Figure 5.3). Although there are a variety of means to categorize flexible fabrics in construction, literature defines three basic types: (1) boundary tensioned membranes, (2) pneumatic or air-based structures, and (3) cable nets or cable beams.[3] The fundamental difference between boundary-tensioned and pneumatic membranes is the method of pre-tensioning. Although both systems possess membrane pre-stresses for stability, boundary-tensioned membranes are mechanically tensioned, while pneumatic structures are tensioned with internal air pressure.

Material innovations for tensile and pneumatic structures have been and continue to be designed to improve fabric stress distribution and air-tightness capabilities. Throughout history, advances in flexible membranes have driven technological and profound sociocultural and economic transformations in architecture.[4] This section provides a brief overview of the fundamental shifts in membrane construction related to material type, whether natural or synthetic, culture, and design teleology.

In 1967, Nikolaus Laing envisioned a breathing membrane. Using a radical method for controlling temperature, ventilation, and humidity transfer through a multivalent air control membrane, Laing presented a parallel vision around polymers and pneumatics (Figure 5.1).[5] Fueled by the desire to create an enclosure that could function as natural organisms' skins, Laing

DOI: 10.4324/9781003613589-6

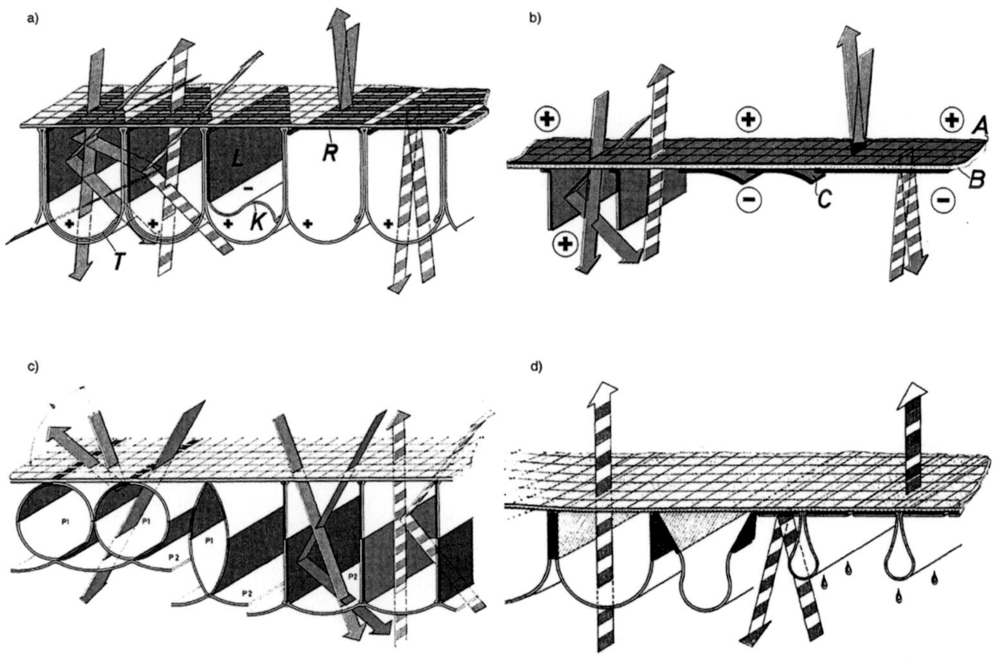

Double-walled pneumatic membrane with a partial metallizedfilm: pressure differentials to augment / control reflective and transmissive, properties (developed by N. Laing, 1967)

FIGURE 5.1 Laing's environmental regulation system for pneumatic membrane structures: (a) pneumatically operated folding film K inside air chamber L regulates between transmission position T and reflection position R; (b) electrostatically controlled design with transparent, conductive outside layer A, an insulating, transparent support layer B, and metallized folding layer C; (c) partly metallized cylindrical air chambers reflect total direct insolation when P1 is inflated or remains highly transmissive when P2 is inflated; and (d) pneumatically controlled design with closed position turning a reflective surface outwards and a black surface inwards which causes strong cooling at night by ground radiation to induce condensation. Diagrams based on N. Laing, "The Use of Solar and Sky Radiation for Air Conditioning of Pneumatic Structures," in IASS, Proceedings of the 1st International Colloquium on Pneumatic Structures, University of Stuttgart, Stuttgart, Germany, 1967, 11–12.

focused, unlike his contemporaries, on integrative climatic and structural performance rather than form-structure. Laing's vision was unique in its focus on integrating structural efficiency with climatic regulation, where sunlight, ventilation, and heat are simultaneously balanced (Figure 5.1a).

Laing's polymer system proposed a double-walled pneumatic membrane that used pressure differentials to augment the exposure of a reflective and transmissive, movable film that is partially metallized (Figure 5.1b). Essentially anticipating the quasi-responsive ethylene-alt-tetrafluoroethylene (ETFE) cushion systems of contemporary buildings, he proposed insulation regulation through adjustments to the degree of inflation. The aperture and enclosure of pneumatic valves created differentiations in air temperature. These induced air movements were managed by exposing a reflective interior surface and a black exterior body, which triggered light water condensation along the interior (Figure 5.1c, 1d). Although reliant upon a mechanical system to pneumatically activate the system, Laing has brought awareness to the potential of controlling a building's environment through an elastic, pneumatic membrane.[6] Furthermore, Laing's membrane regains relevance as architects challenge envelopes that perform as hermetic barriers.[7] Over fifty years after Laing's "breathing" membrane, architects continue to revisit the potential to develop membranes that can be "alive" for self-regulation, not too different from Laing's vision. Can Engineered Living Materials (ELMs) finally be the channel to crystallize this vision?

ELMs is a nascent frontier that harnesses broad interest in science and engineering. The field encompasses the synchronization of classical materials science, microbiology, and synthetic biology for the engineering and augmentation of living systems with pioneering functionalities. ELMs comprise complex biological materials in non-living components.[8] Engineering involves calibrating complex responses to environmental stimuli or being genetically engineered to allow user control over responses, integrating various inputs and functionalities. Research has focused on ELMs, top-down or bottom-up approaches for materials primarily with single functionalities and for small-scale applications, with a few exceptions such as in biocementitious materials.[9] This chapter examines the architectural implications of bottom-up and top-down approaches in current ELMs, with a focus on biopolymers and indoor spaces. Specifically, the text explores how the frontier of ELMs can potentially usher in new understanding of the materialization and conceptualization of breathing membranes in design.

The chapter is organized in three parts: (1) *Introduction*; (2) *Breathing Membranes, Natural Polymers Disruptions*; (3) *ELMs and Future Membranes*; (4) *Lichen-laden hydrogel Biowall System*. Part III serves as a framework for addressing the potential of ELMs in achieving future *breathing enclosures* with integrative architectural functions.

The terms biopolymers, natural biopolymers, synthetic polymers, and plastics used in this text must be defined for text readability. Synthetic and natural polymers are part of the overarching polymer family. Synthetic polymers are those artificially produced in a lab, originating from natural compounds *or* crude oil and gas.[10] Biopolymers produced in a lab are derived from organisms, such as polylactic acid (PLA), and are defined as synthetic biopolymers. Natural biopolymers are natural products, including plants, animals, bacteria, fungi, yeast, or consortia, which refer to multiple organisms such as lichens. Biopolymers are, for the most part, biodegradable. However, not all biopolymers are biodegradable, given that they decompose into compounds such as carbon dioxide, water, and methane by enzymatic action, which depends on the structure and properties of the biopolymer. Hence, there are synthetic biopolymers that are biodegradable and some that are not, such as bio-polyethene.[11] Polymers are also typically referred to as plastics. However, polymers encompass three subfamilies: thermoplastics, thermosets (e.g., epoxy), and elastomers. Hence, the term plastic only refers to thermoplastics in this text.

Natural Polymers Disruptions in Enclosures

Nineteenth-century advances in physics and the development of synthetic polymers streamlined a myriad of innovations in construction, and with it disrupting centuries of use of natural polymers (Figure 5.2, left). From elastic membranes with load-bearing capacity in lightweight membranes to rigid thermoplastics such as polycarbonate and polymethyl methacrylate,[12] a

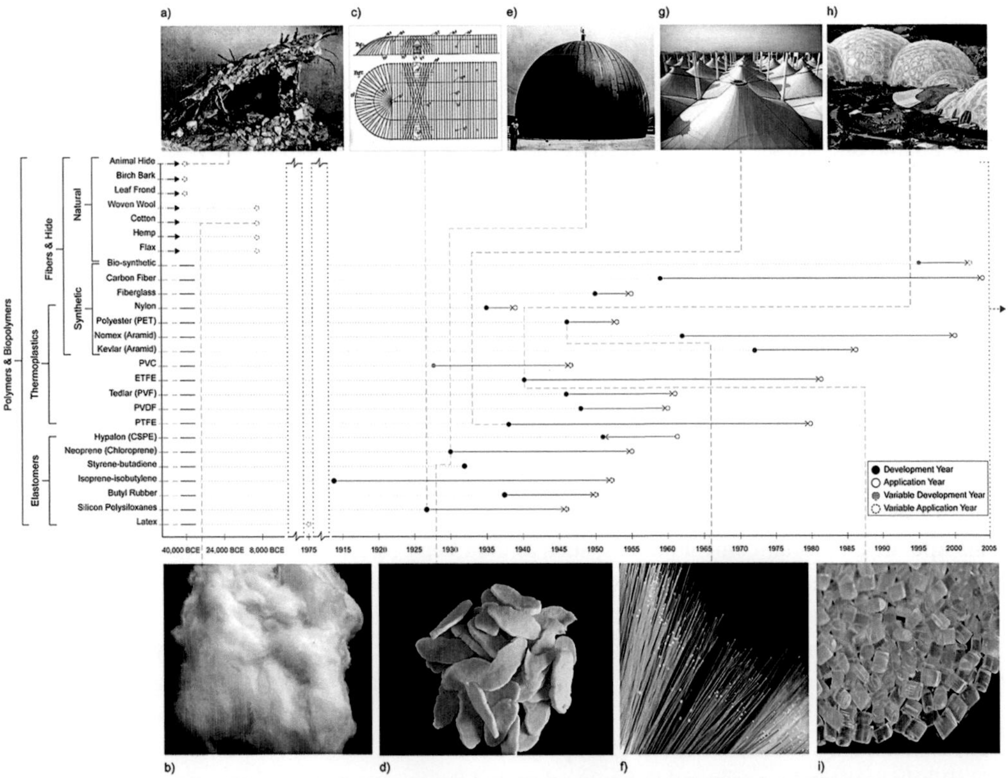

FIGURE 5.2 Polymers and Construction [developments and adaptation]. Summary diagram presenting the relation between material developments (rubbers and fibers) before and after the rise of synthetics and their implementation in architecture: (a) reconstructed hut composed by animal hide and bones ("Stone Age Habitats," accessed November 7, 2013, www.afghanchamber.com/history/stoneages.htm); (b) cotton; (c) field-hospital tent by William Lanchester proposed in oiled silk/natural rubber (Frederick William Lanchester, "Construction of Tents for Field-Hospitals, Depots, and like Purposes," April 29, 1919, https://patents.google.com/patent/US1302182A/en); (d) chloroprene; (e) Radomes developed by Walter Bird ("368c.jpg (JPEG Image, 467 × 350 Pixels)," accessed November 7, 2013, https://findingaids.lib.buffalo.edu/repositories/2/archival_objects/40658); (f) fiberglass; (g) Hajj Terminal designed by Fazlur Khan from Skidmore, Owings & Merrill in glass fiber coated with ETFE ("Hajj_terminal.jpg (JPEG Image, 500 × 375 Pixels)," accessed November 7, 2013, http://upload.wikimedia.org/wikipedia/commons/d/d5/Hajj_terminal.jpg); (h) Eden Project by Nicholas Grimshaw in ETFE ("The_Eden_Project_-_geograph.org.uk_-_217614.jpg (JPEG Image, 640 × 480 Pixels)," accessed November 7, 2013, http://upload.wikimedia.org/wikipedia/commons/3/38/The_Eden_Project_-_geograph.org.uk_-_217614.jpg); and (i) ETFE. (Diagram by author).

shift through the use of synthetic polymers marked the twentieth-century industrialized material culture's predominance in architecture.

The first plastics, such as celluloid and Bakelite developed in the nineteenth century, were introduced to a broader audience at the beginning of the twentieth century as an accessible and sustainable alternative to natural resources.[13] By the mid-twentieth century, it was advertised as colorful, lightweight, and low-priced. It was considered a landmark of its time.[14] From industry labs, such as DuPont, to designers, a strong agenda of promoting, in various forms, the transformational role of plastics in design and society[15] catapulted its integration in architecture. However, our relationships with thermoplastics have been complex, comprised of shifting and varied different perspectives across their history.[16] The history of thermoplastics and elastomers consumption from a structural perspective in architecture has been less turbulent, mainly due to their role in accomplishing portable and lightweight systems.

Tensile and air-based architecture incorporated material inventions at varying speeds (Figure 5.2) across the twentieth century. Polyethene terephthalate, known as PET, was developed in 1941 and incorporated into architecture in 1953. However, many other inventions took decades to become part of the mainstream of architectural design. This is the case of ETFE (1949), which was first incorporated in 1982. Iconic buildings, such as the Eden Project by Grimshaw (2001) (Figure 2h, 2i) and Herzog & de Meuron's Munich Arena (2003), hailed ETFE as the medium for innovation. However, this thermoplastic was initially developed in original polymer form in 1948 (PTFE—Polytetrafluoroethylene) and as ETFE (Ethylene Tetrafluoroethylene) in 1970 by DuPont. Due to its recognized superior durability, strength, and translucency, architects and engineers favored its incorporation in constructions.[17,18] ETFE became hailed as the most structurally and energy efficient not only of elastomers but of all polymers.[19]

While cost can be attributed to having deterred the incorporation of polymers such as ETFE earlier in architecture, as in other material developments, other factors take precedence. The distance between the architects and the material lab invention across the twentieth century[20] have and continue to contribute to this pattern of delayed integration (Figure 5.3) in a broad range of projects since the mid-twentieth century.

Biopolymers, ELMs, and Enclosures

Synthetic polymers still comprise our current material palette for lightweight membranes. When and how are this material trends bound to change? Are recent interests in bioeconomies bound to fundamentally shift not only the material selection but also our understanding of material performance in lightweight architecture this century? If so, what role do the fundamental shifts that have taken place in material science and bioengineering hold in changing these material trends in architecture? To address these inquiries, we must first look into the role of high-performance/smart polymers in construction enclosures of the past twenty years.

The first decade of this century was marked by an interest in technology transfer from the robotics and aerospace industries, including for sensing and actuation capabilities, as a means to test ephemeral boundaries and shape a potentially new frontier in the history of rubber.[21] Synthetic rubbers were used as a cover of prisms, as in the *Hot Box* by McLeod/Croft (2010) and the *Rubber House Pavilion* by Van Gelderen (2010) deliberately to express skins rather than structural systems. These experiments marked a period where rubber was not only harbored in air-form geometries, but as a literal flexible envelope. During this period, we witnessed a shift towards exploring methods to integrate robotics and chemical treatments to transform

FIGURE 5.3 3D hydrogel extrusion. Measurement of reactiveness and morphological deformation/expansion—Detailed View of 3d Elastomeric Printed with Hydrogels Immersed in Water—UC Berkeley Spring Arch 269 Seminar 2015 (MP Gutierrez instructor). Study and photograph by graduate student J. Kommen).

air-based structures. In tandem, new tensile hybrid surface technologies coupled with structural innovations such as Tensairity, which combines tensegrity and pneumatics[22] continued to be developed. Nevertheless, the development of this frontier, as in the use of thermoplastics, has been marked by investigations where the multi-objective performance criteria have been significantly complex and arduous to accomplish.

Reactive materials can be of biological, synthetic, or biosynthetic origin, responding in various formats to different species of external stimuli. Based on the type of outputs, these materials can be either (1) reactive or (2) generative. Single responses in materials are typically classified as mechanical, chemical, electric, optic, magnetic, or thermal. Within each of these general categories, there are subsets (e.g., heat reactive, pH reactive, etc.). Natural materials with reactive properties are typically biopolymers, such as hydrogels (Figure 5.3), derived from alginates sourced from algae or animal sources (collagen). Biomaterials with sensing capabilities can also be of polymeric and mineral combinations.[23] In bulk, reactive synthetic materials are either polymers or polymeric composites. Such advances have been integrated into architecture, including in the form of shape-memory alloys, elastomeric electroactive facades, and hygro- or thermo-responsive flexible panels or membranes.[24]

Scientists have made significant progress in synthetic and biosynthetic reactive materials for flexible membranes over the past twenty years, enabling them to adapt and respond to increasingly complex inputs, including biological signals. For example, nanocomposites enable rapid and tunable motions controlled by the position, intensity, and path of light. For the built environment, such materials have begun to open unprecedented opportunities for efficiency, as building enclosures must adapt and calibrate multiple operations. Laing's aspiration to create a breathing skin, the quintessential expression of a live organism through a mechanistic model over fifty years ago, attempted to be multifunctional. Could it be that materials in which integrated functions can provide more advantages than the mere addition of multiple functions bring his vision into fruition and revolutionize our understanding of enclosure and adaptability?

Explorations in breathing capacities for architecture have led to developments such as artificial stoma as a breathing membrane, reimagining the integration of multifunctional biopolymers (hydrogels) for architectural applications.[25] But how could we move beyond responsiveness, encompassing multifunctionalities that not only emulate living organisms but are literally live matter? Are metabolic activities the truly transformational path to achieving comfort and health in future spaces? This is where the frontier of ELMs can shed new light. The project discussed next stems from this motivation.

Intending to incorporate live matter in construction, architects, engineers, and scientists have turned to the promise of ELMs. This field draws on the fundamental traits of living cells to create functionalized matter that embodies one or various characteristics of biological systems. Scientists draw from a breadth of chassis cells in synthesizing and assembling capacities of hierarchically organized materials with complex structures and outstanding performance.[26] From microorganisms such as algae and fungi to the larger and more complex organisms in plants and animals, research in ELMs is a rapidly growing field[27] including through the incorporation of artificial intelligence for accelerating biosynthesis innovation.[28]

Advances in capabilities to manufacture, such as multiple bioplotting[29] and synthesizing capability at the frontier of ELMs are forging new paths to make this vision possible in the next decade. Scientists and engineers are centering on finding paths to overcome key challenges in our ability to genetically and metabolically engineer cells to produce the designed molecules, scalability, material lifespan, and stabilization.[30] Various taxonomies have been proposed for ELMs, ranging from scale–properties–organism type–design–application domain[31] to those related to life taxonomies instead of materials science disciplines.[32]

While architects have been interested in live matter—be it algae for facades (e.g., BIQ building, Hamburg, Arup), the Urban Algae Farm prototype (Cloud Collective) for at least the past two decades, algae facade prototypes (Algatecture by Carlo Ratti), algae reactor installations (e.g., H.O.R.T.U.S. by ecoLogicStudio), and plant-synthetic structures (e.g., F. Ludwig, D. Schoenle, Baubotanik, FutureArchitecture), and components (e.g., mycelium bricks)—applications of ELMs in the built environment remain at a conceptual level due to scalability challenges and durability.[33] For example, mycelium bricks, while stemming from living material (fungi), undergo a final inactivation process to heat-kill the living cells.[34]

More recently, scientists and architects have begun collaborating to advance the manufacturing capacities of ELMs, both in bioceramics[35] and biopolymers focused on 3D scaffolds[36] and exploring implications of this frontier, including ethics.[37] It is within this context that the *Lichen Blocks* project is positioned.

Lichen-Laden Hydrogels: The Lichen Biowall System

Although research has advanced recently in understanding the chemistry and implications of some indoor air particle sources and the correlation between humidity and air quality, the actual simultaneous removal of VOCs and CO_2 fixation and maintenance of hygrothermal comfort remains largely unaccomplished.[38] In comparison to urban atmospheres, indoor environments feature large surface-area-to-volume ratios.[39] A primary non-reactive species emitted into an indoor atmosphere is approximately 1000 times more likely to be inhaled than if those same species were emitted into the outdoor air.[40] Consequently, advancing our capacity to remove indoor air contaminants is a highly complex challenge that involves multiple disciplines, including health, science, engineering, and architecture. Lichens' primary distinctive features related to indoor air quality can be summarized as follows: humidity control, high surface-to-volume ratio, survivability, carbon sequestration, VOC uptake, and growth on lignocellulose.

Humidity & Surface-Ratio—Lichen Blocks research operates at this interdisciplinary intersection, exploring the opportunity to work with an organism that can potentially tackle the challenge of large surface-area-to-volume ratios and humidity control in indoor spaces. It seeks to shift radically from biowalls made of plants, which have proven to have high limitations in de facto detoxification and energy efficiency.[41] Since lichens do not produce evapotranspiration like plants, nor do they require soil to grow, and they do not have stomatal blockage, they are highly promising as sources of regenerative materials for the built environment.

Living Consortia—Lichens are complex symbiotic organisms of yeast, algae, fungi, and bacteria.[42] In lichens, algae produce carbon compounds through photosynthesis. The fungal partner, in return, provides protection and collects nutrients and moisture from the environment.[43] These organisms have distinctive survivability[44] and have a unique role in the atmospheric fixation of carbon dioxide and nitrogen cycles.[45] Lichens' resilience shows survivability even in simulated Martian environmental conditions.[46]

Detoxification and Carbon Sequestration—Lichens are excellent accumulators of macro and microelements, making them exceptional biomonitors of pollutants, including indoor pollution.[47] Lichens have demonstrated the capacity to capture various pollutants through their cortex (porous surface at the top of lichen thalli or body) and their root-like (rhizine)[48] including toxins from metals.[49] Lichens have shown the capacity to capture pollutants, including organic contaminants such as polycyclic aromatic hydrocarbons[50] and volatile organic compounds uptake through complex processes of metabolizing nutrients.[51] Lichens are also excellent carbon dioxide assimilators. Lichens use the carbon dioxide fixed through photosynthesis to form biomass.[52]

Lichen Growth on Biomass—Lichens grow on an array of varying different substrates from rocks to woods.[53] These organisms colonize trees at varying rates, dependent on factors such as surface roughness, chemical composition, and environmental conditions (e.g., moisture, light, texture, and pH levels).[54] This project explores how to synergize the lichen growth and surface ratio of lichens onto barks with optimized 3D structures.

The *Lichen Blocks* project investigates how to enhance the carbon sink capacity of bark residues and VOC uptake in an ultralight structural system. Specifically, the research explores the synergies of extruding lichens within hydrogels to create 3D scaffolds with optimized geometries onto lignocellulose (bark residues) for indoor air detoxification and humidity control purposes. To date this research has focused on evaluating the (i) manufacturing viability to create 3D scaffolds that are easily scalable for architectural aims; (ii) comparative analysis of the role of relative humidity on the growth and photosynthetic activity of dry lichen versus lichen-laden hydrogels under varying timescales from six weeks to one year; (iii) adherence of the lichen-laden hydrogel onto extruded recycled barks and PLA composites; (iv) carbon sink

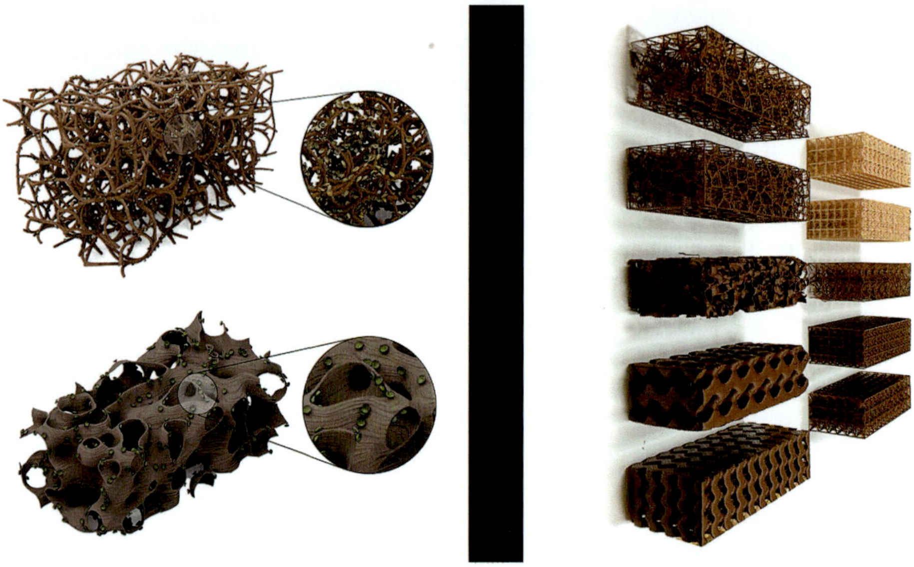

FIGURE 5.4 Left: Lichen block design for surface-ratio augmentation and lightweight structural capacities. Right: Lignocellulosic 3D scaffold with varying design optimizations of lightweight structural performance and surface-ratio maximization (by author, 2023).

capacity of the biowall system. These studies have been conducted iteratively across varying geometries and time spans.

Unlike algae, lichen has not been engineered or encapsulated into hydrogels to date. Lichen Blocks' initial investigations focused on the design and manufacturing of 2D and 3D extrusion of lichen-laden hydrogels onto lignocellulose and maximization of surface ratio for reactivity (Figure 5.4), paving the way for future scalability and geometric control needs. This included exploring effective encapsulation protocols through bioplotting (Figure 5.5) and optimizing the robotically controlled lignocellulosic (bark) scaffolds (Figure 5.6). In the study, an assessment was conducted of various factors, including angle variation, lichen concentration, the number of scaffold layers, uneven and discontinuous outputs, and differences based on the hydrogel's crosslinking.

The study has also focused on evaluating the capacity of lichen-laden hydrogels to maintain continued metabolic activity, coupled with morphological and structural functionalities. To achieve this, we have conducted multiple imaging analyses of metabolic activity. Imaging analysis including Confocal fluorescence of the lichen-laden hydrogel (Figure 5.7) scaffold demonstrating continued metabolic activity after six months without hydration. This result is particularly relevant for the feasibility of incorporating a biowall system that is maintenance-free and requires no additional infrastructure to operate, unlike current biowalls. However, there is a dimension that is as relevant as the technological implications of lichen-laden biowalls, which pertains to the social, psychological, and ethical dimensions of ELMs in indoor spaces. As with any living organism in an indoor space, we must not only guarantee a sustainable harvesting and production but also investigate the implications of what it is to live in blurred boundaries of abiotic and living matter.

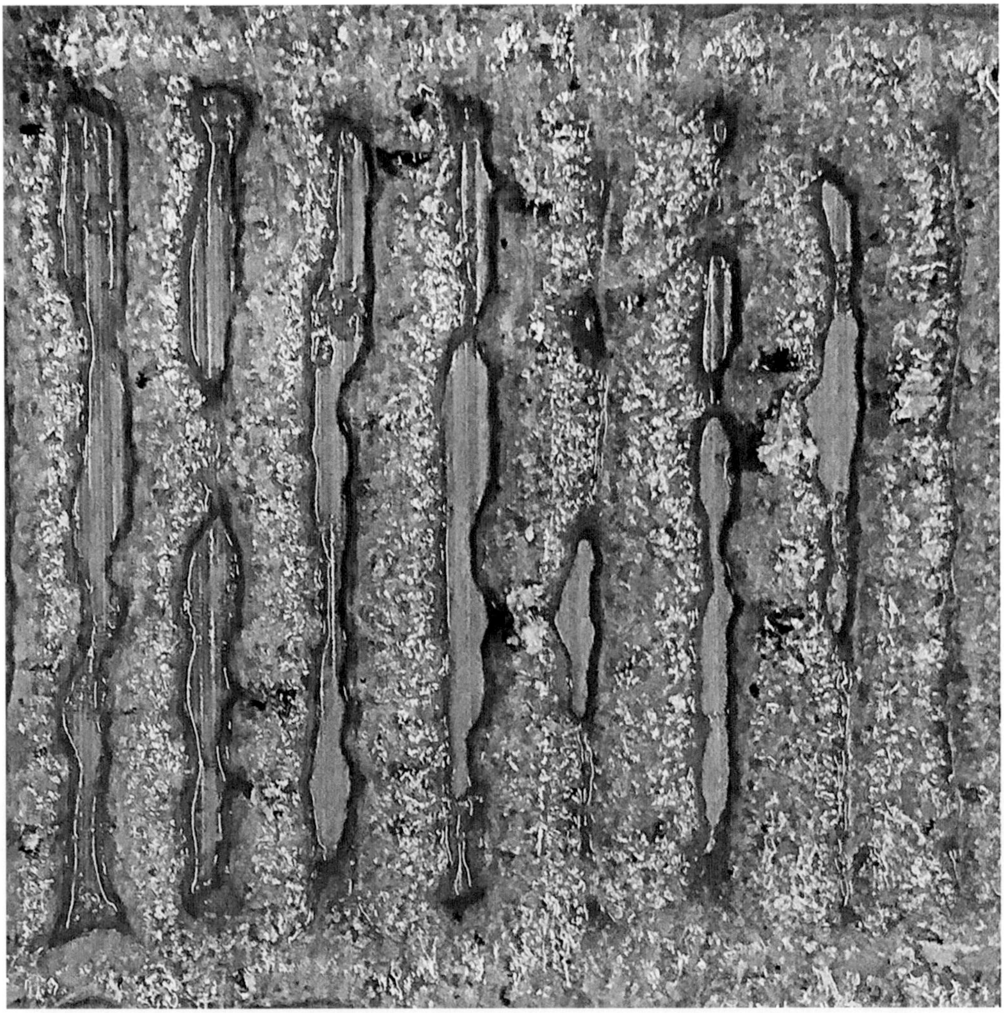

FIGURE 5.5 Lichen-laden hydrogel design and fabrication- A: Lichen-laden hydrogel extrusion with high viscosity onto cork-PLA-PHA substrate 0°raster angle (by author, 2023).

The ongoing explorations of *Lichen Blocks* have focused on incorporating additional bark extractives, such as pine resin, to explore additional functionalities, including antimicrobial properties (Figure 5.8). Extractives of barks, including suberin in cork,[55] have been broadly studied in their natural form as essential to the defense mechanisms of trees. However, the role of extractives in enclosures made of forestry residues for construction remains untapped. Advancing antimicrobial capacities of enclosures holds great promise for ELMs in future construction, both interior and exterior applications. The Lichen blocks project requires future extensive research in this area from both manufacturing and laboratory (biosynthesis) perspectives.

FIGURE 5.6 Lichen-laden hydrogel onto 3D lignocellulosic scaffold robotically controlled; Left (spinning of lichen-laden hydrogel; Right: extrusion onto lignocellulosic scaffold (by author, 2024).

Forethoughts

The use of polymers over the past century in construction has significantly streamlined the technological leaps in lightweight construction. Through this process, new opportunities for adaptability and ephemerality in architecture arose, with profound social implications. In tandem, the expansion of primarily synthetic materials has also heavily implicated these technological processes in the generation of waste that impacts the environment for thousands of years. It is within this reality that architects and engineers have long envisioned a breathing membrane that approximates the behavior of natural skin. Over fifty years ago, this vision was epitomized by Laing, stemming from a mechanistic view. In this century, new material capabilities have emerged to make this vision feasible, veering into reactive polymers with sensing and actuation capabilities. However, are we moving beyond these aims and searching for systems that are not only responsive and adaptive, but also bear functionalities with the evolving intelligence of live organisms? Is this bound to transform architectural enclosures in the next decade?

Materials with programmable regeneration functionalities hold distinctive promises for meeting the ever-growing demands of risk resilience and efficacy in constructions. Given this need, the integration of ELMs in architecture can potentially transform our capacity to achieve this aim. Blurring boundaries between abiotic and living matter in future enclosure technologies with tangible breathing functions is not a purely technological pursuit. As we enter this realm, it is crucial to recognize that the evolving distinctions between living and non-living entities in construction are rooted in the interplay of science, ethics, society, and design.

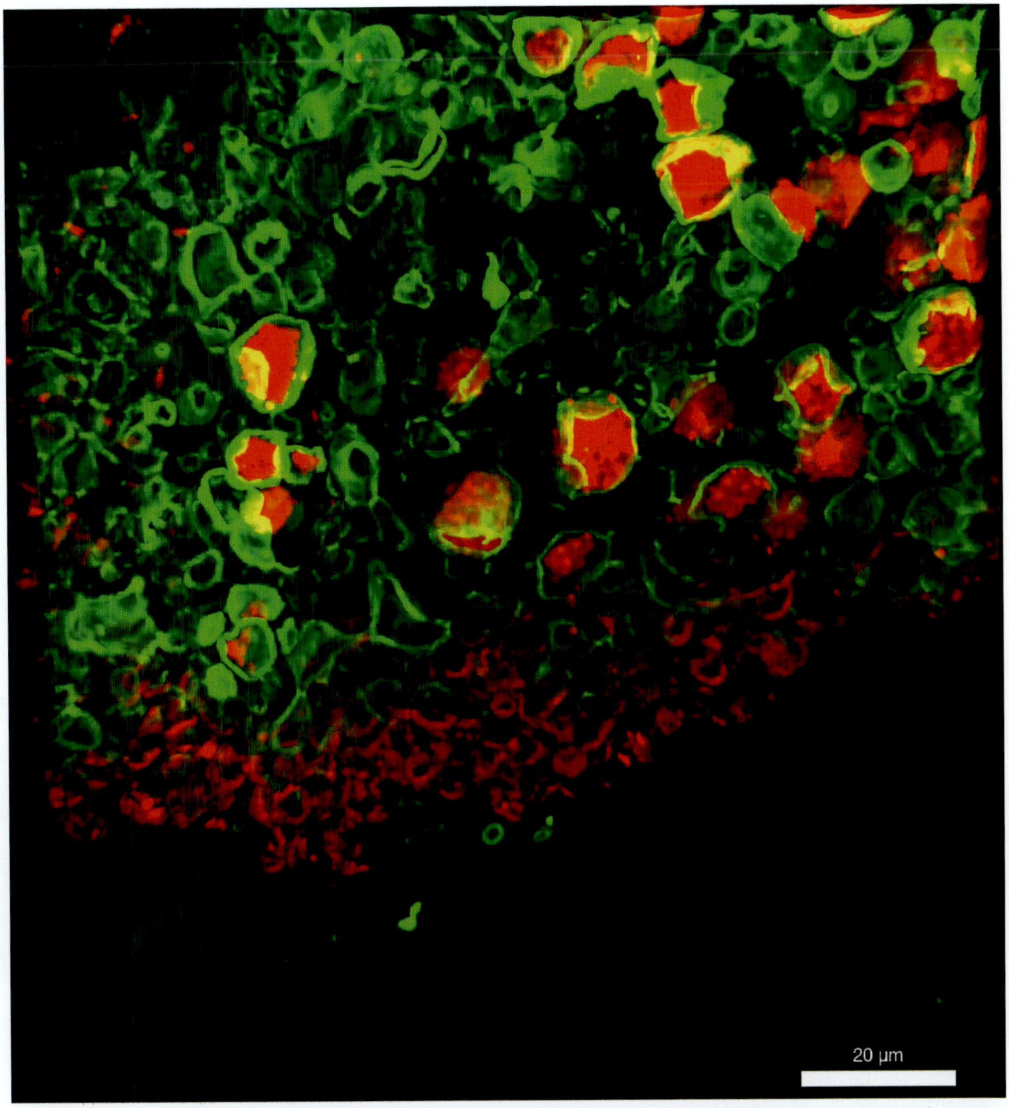

FIGURE 5.7 Confocal fluorescence image of lichen-laden hydrogel scaffold fragment. Green is calcofluor white staining of fungal hyphae, red is chlorophyll autofluorescence (by author in collaboration with Dr. Klara Sharnagl, 2023). Image displayed at the 2025 Architecture Venice Biennale (author MP Gutierrez invited).

Acknowledgments

The author is grateful for the research support of the Bakar Prize 2024.

FIGURE 5.8 Detailed view of lichen block (Phase 2). Non-planar granular bark extrusion of lichen blocks. The block is composed of lichen-laden hydrogels, lignocellulose with the incorporation of bark extractives (pine resin) surfacing for antimicrobial functionalities (by author, 2024).

Notes

1 Gottfried Semper, *The four elements of architecture and other writings* (Cambridge University Press, 1989).
2 Maria Paz Gutierrez, "Programming Selective Elasticity," in *Green Design, Materials and Manufacturing Processes*, ed. Helena Bartolo et al. (CRC PressI Llc, 2013).
3 W. J Lewis, *Tension Structures: Form and Behaviour* (Thomas Telford, 2003); Ronald E. Shaeffer, *Tensioned Fabric Structures: A Practical Introduction* (The Society, 1996).
4 Hadas Steiner, "The Forces of Matter," *The Journal of Architecture* 10, no. 1 (2005): 91–109.
5 Universität Stuttgart (TH) Department of Architecture et al., *Proceedings of the 1st International Colloquium on Pneumatic Structures, May 11 and 12, 1967, University of Stuttgart, Germany* (University, IASS, 1967).

6 N. Laing, "The Use of Solar and Sky Radiation for Air Conditioning of Pneumatic Structures," *IASS, Proceedings of the 1st International Colloquium on Pneumatic Structures, University of Stuttgart, Stuttgart, Germany*, 1967, 11–12.

7 Annette W. LeCuyer et al., *ETFE: Technology and Design* (Birkhäuser, 2008); Michelle Addington, "Contingent Behaviours," *Architectural Design* 79, no. 3 (2009): 12–17.

8 Aleixandre Rodrigo-Navarro et al., "Engineered Living Biomaterials," *Nature Reviews Materials* 6, no. 12 (2021): 1175–90, https://doi.org/10.1038/s41578-021-00350-8.

9 Hsiao Wei Lee et al., "Prediction of Microbial-Induced Calcium Carbonate Precipitation for Self-Healing Cementitious Material," *Cement and Concrete Composites* 158 (April 2025): 105945, https://doi.org/10.1016/j.cemconcomp.2025.105945; Danielle N. Beatty and Wil V. Srubar, "Nucleation Effects of Coccoliths in Portland Cement," *Matter*, April 2025, 102100, https://doi.org/10.1016/j.matt.2025.102100.

10 Galina Satchanska et al., "Natural and Synthetic Polymers for Biomedical and Environmental Applications," *Polymers* 16, no. 8 (2024): 1159, https://doi.org/10.3390/polym16081159.

11 Aya Samir et al., "Recent Advances in Biodegradable Polymers for Sustainable Applications," *Npj Materials Degradation* 6, no. 1 (2022): 68, https://doi.org/10.1038/s41529-022-00277-7.

12 Maria Paz Gutierrez and Tarek I. Zohdi, "Effective Reflectivity and Heat Generation in Sucrose and PMMA Mixtures," *Energy and Buildings* 71 (March 2014): 95–103, https://doi.org/10.1016/j.enbuild.2013.11.046.

13 Simone Jeska, *Transparent Plastics: Design and Technology* (Springer, 2008), http://books.google.com/books?hl=en&lr=&id=0OML58qoE0IC&oi=fnd&pg=PA7&dq=Transparent+Plastics+Design+and+Technology&ots=GDi8rl7VTA&sig=FWpPZ_56fROeC9hSja5joUGB4aE.

14 Cristina Bargna, "Synthetic Paradoxes: Moving with the Dematerialization of Plastic Design," *Journal of Design History*, May 21, 2025, epaf003, https://doi.org/10.1093/jdh/epaf003.

15 Jeffrey L. Meikle, *American Plastic: A Cultural History* (Rutgers University Press, 1997).

16 Louise Dennis, "A Brief History of the Use of Plastics," *Cambridge Prisms: Plastics* 2 (2024): e19, https://doi.org/10.1017/plc.2024.17; Meikle, *American Plastic*.

17 LeCuyer et al., *ETFE*.

18 Stefanie Reese, "Meso-Macro Modelling of Fibre-Reinforced Rubber-like Composites Exhibiting Large Elastoplastic Deformation," *International Journal of Solids and Structures* 40, no. 4 (2003): 951–80.

19 Dong Yang Wu et al., "Self-Healing Polymeric Materials: A Review of Recent Developments," *Progress in Polymer Science* 33, no. 5 (2008): 479–522, www.sciencedirect.com/science/article/pii/S0079670008000208; S Robinson-Gayle, "ETFE foil cushions in roofs and atria," *Construction and Building Materials* 15, no. 7 (2001): 323–27; John Fernandez, *Material Architecture* (Taylor and Francis, 2012).

20 Maria-Paz Gutierrez, "Reorienting Innovation: Transdisciplinary Research and Building Technology," *Architectural Research Quarterly* 18, no. 01 (2014): 69–82, https://doi.org/10.1017/S1359135514000372.

21 Colin Barlow et al., *The World Rubber Industry* (Routledge, 1994); John A Tully, *The Devil's Milk: A Social History of Rubber* (Monthly Review Press, 2011).

22 Martin Bechthold, *Innovative Surface Structures: Technology and Applications* (Taylor & Francis, 2008); Axel Ritter, *Smart Materials: In Architecture, Interior Architecture and Design* (Springer, 2007); Steven Van Dessel and Abdol R. Chini, "Rigidified Pneumatic Composites: Use of Space Technologies to Build the next Generation of American Homes," National Science Foundation Award Abstract #0122022, n.d.

23 Xiaoxing Xia et al., "Responsive Materials Architected in Space and Time," *Nature Reviews Materials* 7, no. 9 (2022): 683–701; Joanna Aizenberg and Peter Fratzl, "New Materials through Bioinspiration and Nanoscience," *Advanced Functional Materials* 23, no. 36 (2013): 4398–99; Peter Fratzl and Friedrich G. Barth, "Biomaterial Systems for Mechanosensing and Actuation," *Nature* 462, no. 7272 (2009): 442–48; Philip Ball, *Shapes: Nature's Patterns: A Tapestry in Three Parts*, vol. 1 (Oxford University Press, 2011).

24 Martin Bechthold and James C. Weaver, "Materials Science and Architecture," *Nature Reviews Materials* 2, no. 12 (2017): 17082, https://doi.org/10.1038/natrevmats.2017.82.

25 Younggeun Park et al., "Reversible Self-Actuated Thermo-Responsive Pore Membrane," *Scientific Reports* 6, no. 1 (2016), https://doi.org/10.1038/srep39402.

26 Yanyi Wang et al., "Engineered Living Materials (ELMs) Design: From Function Allocation to Dynamic Behavior Modulation," *Current Opinion in Chemical Biology* 70 (October 2022): 102188, https://doi.org/10.1016/j.cbpa.2022.102188.
27 Yonatan Chemla et al., "Design and Regulation of Engineered Bacteria for Environmental Release," *Nature Microbiology* 10, no. 2 (2025): 281–300, https://doi.org/10.1038/s41564-024-01918-0; Rodrigo-Navarro et al., "Engineered Living Biomaterials"; O. Burgos-Morales et al., "Synthetic Biology as Driver for the Biologization of Materials Sciences," *Materials Today Bio* 11 (June 2021): 100115, https://doi.org/10.1016/j.mtbio.2021.100115; Tzu-Chieh Tang et al., "Materials Design by Synthetic Biology," *Nature Reviews Materials* 6, no. 4 (2021): 332–50, https://doi.org/10.1038/s41578-020-00265-w; Wil V. Srubar, "Engineered Living Materials: Taxonomies and Emerging Trends," *Trends in Biotechnology* 39, no. 6 (2021): 574–83, https://doi.org/10.1016/j.tibtech.2020.10.009; Lina M. González et al., "Resilient Living Materials Built by Printing Bacterial Spores," *Nature Chemical Biology* 16, no. 2 (2020): 126–33, https://doi.org/10.1038/s41589-019-0412-5; Chemla et al., "Design and Regulation of Engineered Bacteria for Environmental Release."
28 M.P. Gutierrez, "Material Alignments," *Technology|Architecture + Design* 5, no. 1 (2021): 115–18, https://doi.org/10.1080/24751448.2021.1887682.
29 Jonathan D. Weiss et al., "A Low-Cost, Open-Source 3D Printer for Multimaterial and High-Throughput Direct Ink Writing of Soft and Living Materials," *Advanced Materials* 37, no. 10 (2025): 2414971, https://doi.org/10.1002/adma.202414971.
30 Chemla et al., "Design and Regulation of Engineered Bacteria for Environmental Release"; Srubar, "Engineered Living Materials"; Jonathan D Weiss et al., "A Low-Cost, Open-Source 3D Printer for Multimaterial and High-Throughput Direct Ink Writing of Soft and Living Materials," *Advanced Materials*, Wiley Online Library, 2025, 2414971; Sara Molinari et al., "Bottom-up Approaches to Engineered Living Materials: Challenges and Future Directions," *Matter* 4, no. 10 (2021): 3095–120.
31 Srubar, "Engineered Living Materials."
32 Andrés Díaz Lantada et al., "Taxonomy for Engineered Living Materials," *Cell Reports Physical Science* 3, no. 4 (2022): 100807, https://doi.org/10.1016/j.xcrp.2022.100807.
33 Sara J. Wilkinson et al., "Algae Building: Is This the New Smart Sustainable Technology?," in *Data-Driven Multivalence in the Built Environment*, ed. Nimish Biloria, S.M.A.R.T. Environments (Springer International Publishing, 2020), https://doi.org/10.1007/978-3-030-12180-8_12.
34 Elise Elsacker et al., "Growing Living and Multifunctional Mycelium Composites for Large-Scale Formwork Applications Using Robotic Abrasive Wire-Cutting," *Construction and Building Materials* 283 (May 2021): 122732, https://doi.org/10.1016/j.conbuildmat.2021.122732; Mitchell Jones et al., "Engineered Mycelium Composite Construction Materials from Fungal Biorefineries: A Critical Review," *Materials & Design* 187 (February 2020): 108397, https://doi.org/10.1016/j.matdes.2019.108397.
35 Lee et al., "Prediction of Microbial-Induced Calcium Carbonate Precipitation for Self-Healing Cementitious Material."
36 Israel Kellersztein et al., "Multifunctional Biocomposite Materials from *Chlorella Vulgaris* Microalgae," *Advanced Materials* 37, no. 1 (2025): 2413618, https://doi.org/10.1002/adma.202413618; Anja Lode et al., "Green Bioprinting: Fabrication of Photosynthetic Algae-Laden Hydrogel Scaffolds for Biotechnological and Medical Applications," *Engineering in Life Sciences* 15, no. 2 (2015): 177–83, https://doi.org/10.1002/elsc.201400205.
37 Andrés Díaz Lantada et al., "Blurring the Boundaries between Living and Non-Living? Examining and Mapping the Living Properties of Engineered Matter and Systems," *Technology in Society* 81 (June 2025): 102822, https://doi.org/10.1016/j.techsoc.2025.102822.
38 Yingrui Zhu et al., "A Literature Review Investigating the Impact of Temperature and Humidity on Volatile Organic Compound Emissions from Building Materials," *Building and Environment* 262 (August 2024): 111845, https://doi.org/10.1016/j.buildenv.2024.111845; Peder Wolkoff et al., "Health, Work Performance, and Risk of Infection in Office-like Environments: The Role of Indoor Temperature, Air Humidity, and Ventilation," *International Journal of Hygiene and Environmental Health* 233 (April 2021): 113709, https://doi.org/10.1016/j.ijheh.2021.113709; Peder Wolkoff, "Indoor Air Humidity, Air Quality, and Health – An Overview," *International Journal of Hygiene and Environmental Health* 221, no. 3 (2018): 376–90, https://doi.org/10.1016/j.ijheh.2018.01.015; G. Clausen et al., "Reflections on

the State of Research: Indoor Environmental Quality," *Indoor Air* 21, no. 3 (2011): 219–30, https://doi.org/10.1111/j.1600-0668.2010.00706.x.
39 William W Nazaroff, "INDOOR PARTICLE DYNAMICS," in *Indoor Air*, vol. 14, no. Supplement 7 (2004).
40 Benoit Guieysse et al., "Biological Treatment of Indoor Air for VOC Removal: Potential and Challenges," *Biotechnology Advances* 26, no. 5 (2008): 398–410, https://doi.org/10.1016/j.biotechadv.2008.03.005.
41 Bryan E. Cummings and Michael S. Waring, "Potted Plants Do Not Improve Indoor Air Quality: A Review and Analysis of Reported VOC Removal Efficiencies," *Journal of Exposure Science & Environmental Epidemiology* 30, no. 2 (2020): 253–61, https://doi.org/10.1038/s41370-019-0175-9.
42 Toby Spribille et al., "Evolutionary Biology of Lichen Symbioses," *New Phytologist* 234, no. 5 (2022): 1566–82, https://doi.org/10.1111/nph.18048.
43 Lucia Muggia et al., "Effects of Growth Media on the Diversity of Culturable Fungi from Lichens," *Molecules* 22, no. 5 (2017): 824, https://doi.org/10.3390/molecules22050824; Ines Aline Aschenbrenner et al., "Microbial Cargo: Do Bacteria on Symbiotic Propagules Reinforce the Microbiome of Lichens?: Bacterial Cargo on Lichen Propagules," *Environmental Microbiology* 16, no. 12 (2014): 3743–52, https://doi.org/10.1111/1462-2920.12658.
44 Giuseppina Fanelli et al., "Metabolomics of Dry Versus Reanimated Antarctic Lichen-Dominated Endolithic Communities," *Life* 11, no. 2 (2021): 96, https://doi.org/10.3390/life11020096.
45 Ashitha B. Arun et al., "Photobiont Symbiotic Association in Lichens," in *Microbial Symbionts* (Elsevier, 2023), https://doi.org/10.1016/B978-0-323-99334-0.00009-8.
46 Christian Lorenz et al., "Survivability of the Lichen Xanthoria Parietina in Simulated Martian Environmental Conditions," *Scientific Reports* 13, no. 1 (2023): 4893, https://doi.org/10.1038/s41598-023-32008-6.
47 Luca Paoli et al., "Lichens and Mosses as Biomonitors of Indoor Pollution," *Biology* 12, no. 9 (2023): 1248, https://doi.org/10.3390/biology12091248; Nuno Canha et al., "Contribution of Short Irradiation Instrumental Neutron Activation Analysis to Assess Air Pollution at Indoor and Outdoor Environments Using Transplanted Lichens," *Journal of Radioanalytical and Nuclear Chemistry* 320, no. 1 (2019): 129–37, https://doi.org/10.1007/s10967-019-06461-5.
48 Kristin Palmqvist et al., "The Carbon Economy of Lichens," in *Lichen Biology, Second Edition* (Cambridge University Press, 2008), https://doi.org/10.1017/CBO9780511790478.011.
49 Ankita H. Tripathi et al., "Lichens as Bioremediation Agents—A Review," in *Synergistic Approaches for Bioremediation of Environmental Pollutants: Recent Advances and Challenges* (Elsevier, 2022), https://doi.org/10.1016/B978-0-323-91860-2.00015-4.
50 María Blasco et al., "Behaviour of Different Lichen Species as Biomonitors of Air Pollution by PAHs in Natural Ecosystems," *Journal of Environmental Monitoring* 13, no. 9 (2011): 2588–96, https://doi.org/10.1039/c0em00681e; Jill E. Schrlau et al., "Comparison of Lichen, Conifer Needles, Passive Air Sampling Devices, and Snowpack as Passive Sampling Media to Measure Semi-Volatile Organic Compounds in Remote Atmospheres," *Environmental Science & Technology* 45, no. 24 (2011): 10354–61, https://doi.org/10.1021/es202418f.
51 Achim Edtbauer et al., "Cryptogamic Organisms Are a Substantial Source and Sink for Volatile Organic Compounds in the Amazon Region," *Communications Earth & Environment* 2, no. 1 (2021): 258, https://doi.org/10.1038/s43247-021-00328-y.
52 Palmqvist et al., "The Carbon Economy of Lichens."
53 Patrick Jung et al., "Lichens Bite the Dust – A Bioweathering Scenario in the Atacama Desert," *iScience* 23, no. 11 (2020): 101647, https://doi.org/10.1016/j.isci.2020.101647; T. G. A. Green et al., "Physiological Ecology of Carbon Dioxide Exchange," in *Lichen Biology*, 2nd ed., ed. Thomas H. Nash (Cambridge University Press, 2008), https://doi.org/10.1017/CBO9780511790478.010.
54 Toby Spribille et al., "Lichens on Dead Wood: Species-Substrate Relationships in the Epiphytic Lichen Floras of the Pacific Northwest and Fennoscandia," *Ecography* 31, no. 6 (2008): 741–50, https://doi.org/10.1111/j.1600-0587.2008.05503.x; Irwin M. Brodo, "SUBSTRATE ECOLOGY," in *The Lichens* (Elsevier, 1973), https://doi.org/10.1016/B978-0-12-044950-7.50017-9.
55 Filipa Gonçalves et al., "Evaluation of Antimicrobial Properties of Cork," *FEMS Microbiology Letters* 363, no. 3 (2016): fnv231, https://doi.org/10.1093/femsle/fnv231.

References

Addington, Michelle. "Contingent Behaviours." *Architectural Design* 79, no. 3 (2009): 12–17.

Aizenberg, Joanna, and Peter Fratzl. "New Materials through Bioinspiration and Nanoscience." *Advanced Functional Materials* 23, no. 36 (2013): 4398–99.

Aleixandre, Rodrigo-Navarro, Shrikrishnan Sankaran, Matthew J. Dalby, Aránzazu del Campo, and Manuel Salmeron-Sanchez. "Engineered Living Biomaterials." *Nature Reviews Materials* 6, no. 12 (2021): 1175–90. https://doi.org/10.1038/s41578-021-00350-8.

Architecture, Universität Stuttgart (TH) Department of, Civil Engineering, and International Association for Shell Structures. *Proceedings of the 1st International Colloquium on Pneumatic Structures, May 11 and 12, 1967,* University of Stuttgart, Germany. University, IASS, 1967.

Arun, Ashitha B., Shabari Girish, and Lokesh Ravi. "Photobiont Symbiotic Association in Lichens." In *Microbial Symbionts*, edited by Dhanasekaran Dharumadurai. Elsevier, 2023. https://doi.org/10.1016/B978-0-323-99334-0.00009-8.

Aschenbrenner, Ines Aline, Massimiliano Cardinale, Gabriele Berg, and Martin Grube. "Microbial Cargo: Do Bacteria on Symbiotic Propagules Reinforce the Microbiome of Lichens?: Bacterial Cargo on Lichen Propagules." *Environmental Microbiology* 16, no. 12 (2014): 3743–52. https://doi.org/10.1111/1462-2920.12658.

Ball, Philip. *Shapes: Nature's Patterns: A Tapestry in Three Parts*. Vol. 1. Oxford University Press, 2011.

Bargna, Cristina. "Synthetic Paradoxes: Moving with the Dematerialization of Plastic Design." *Journal of Design History* 38, (May 21, 2025): epaf003. https://doi.org/10.1093/jdh/epaf003.

Barlow, Colin, S. K Jayasuriya, and C. Suan Tan. *The World Rubber Industry*. Routledge, 1994.

Beatty, Danielle N., and Wil V. Srubar. "Nucleation Effects of Coccoliths in Portland Cement." *Matter* 8, (April 2025): 102100. https://doi.org/10.1016/j.matt.2025.102100.

Bechthold, Martin. *Innovative Surface Structures: Technology and Applications*. Taylor & Francis, 2008.

Bechthold, Martin, and James C. Weaver. "Materials Science and Architecture." *Nature Reviews Materials* 2, no. 12 (2017): 17082. https://doi.org/10.1038/natrevmats.2017.82.

Blasco, María, Celia Domeño, Patricia López, and Cristina Nerín. "Behaviour of Different Lichen Species as Biomonitors of Air Pollution by PAHs in Natural Ecosystems." *Journal of Environmental Monitoring* 13, no. 9 (2011): 2588–96. https://doi.org/10.1039/c0em00681e.

Brodo, Irwin M. "Substrate Ecology." In *The Lichens*, edited by Vernon Ahmadjian and Mason E. Hale. Elsevier, 1973. https://doi.org/10.1016/B978-0-12-044950-7.50017-9.

Burgos-Morales, O., M. Gueye, L. Lacombe, et al. "Synthetic Biology as Driver for the Biologization of Materials Sciences." *Materials Today Bio* 11 (June 2021): 100115. https://doi.org/10.1016/j.mtbio.2021.100115.

Canha, Nuno, Maria Do Carmo Freitas, and Susana Marta Almeida. "Contribution of Short Irradiation Instrumental Neutron Activation Analysis to Assess Air Pollution at Indoor and Outdoor Environments Using Transplanted Lichens." *Journal of Radioanalytical and Nuclear Chemistry* 320, no. 1 (2019): 129–37. https://doi.org/10.1007/s10967-019-06461-5.

Chemla, Yonatan, Connor J. Sweeney, Christopher A. Wozniak, and Christopher A. Voigt. "Design and Regulation of Engineered Bacteria for Environmental Release." *Nature Microbiology* 10, no. 2 (2025): 281–300. https://doi.org/10.1038/s41564-024-01918-0.

Clausen, G., G. Bekö, R. L. Corsi, et al. "Reflections on the State of Research: Indoor Environmental Quality." *Indoor Air* 21, no. 3 (2011): 219–30. https://doi.org/10.1111/j.1600-0668.2010.00706.x.

Cummings, Bryan E., and Michael S. Waring. "Potted Plants Do Not Improve Indoor Air Quality: A Review and Analysis of Reported VOC Removal Efficiencies." *Journal of Exposure Science & Environmental Epidemiology* 30, no. 2 (2020): 253–61. https://doi.org/10.1038/s41370-019-0175-9.

Dennis, Louise. "A Brief History of the Use of Plastics." *Cambridge Prisms: Plastics* 2 (2024): e19. https://doi.org/10.1017/plc.2024.17.

Edtbauer, Achim, Eva Y. Pfannerstill, Ana Paula Pires Florentino, et al. "Cryptogamic Organisms Are a Substantial Source and Sink for Volatile Organic Compounds in the Amazon Region." *Communications Earth & Environment* 2, no. 1 (2021): 258. https://doi.org/10.1038/s43247-021-00328-y.

Elsacker, Elise, Asbjørn Søndergaard, Aurélie Van Wylick, Eveline Peeters, and Lars De Laet. "Growing Living and Multifunctional Mycelium Composites for Large-Scale Formwork Applications Using

Robotic Abrasive Wire-Cutting." *Construction and Building Materials* 283 (May 2021): 122732. https://doi.org/10.1016/j.conbuildmat.2021.122732.

Fanelli, Giuseppina, Claudia Coleine, Federica Gevi, Silvano Onofri, Laura Selbmann, and Anna Maria Timperio. "Metabolomics of Dry Versus Reanimated Antarctic Lichen-Dominated Endolithic Communities." *Life* 11, no. 2 (2021): 96. https://doi.org/10.3390/life11020096.

Fernandez, John. *Material Architecture*. Taylor and Francis, 2012.

Fratzl, Peter, and Friedrich G. Barth. "Biomaterial Systems for Mechanosensing and Actuation." *Nature* 462, no. 7272 (2009): 442–48.

Gonçalves, Filipa, Patrícia Correia, Susana P. Silva, and Cristina Almeida-Aguiar. "Evaluation of Antimicrobial Properties of Cork." *FEMS Microbiology Letters* 363, no. 3 (2016): fnv231. https://doi.org/10.1093/femsle/fnv231.

González, Lina M., Nikita Mukhitov, and Christopher A. Voigt. "Resilient Living Materials Built by Printing Bacterial Spores." *Nature Chemical Biology* 16, no. 2 (2020): 126–33. https://doi.org/10.1038/s41589-019-0412-5.

Green, T. G. A., T. H. Nash, and O. L. Lange. "Physiological Ecology of Carbon Dioxide Exchange." In *Lichen Biology*, 2nd ed., edited by Thomas H. Nash. Cambridge University Press, 2008. https://doi.org/10.1017/CBO9780511790478.010.

Guieysse, Benoit, Cecile Hort, Vincent Platel, Raul Munoz, Michel Ondarts, and Sergio Revah. "Biological Treatment of Indoor Air for VOC Removal: Potential and Challenges." *Biotechnology Advances* 26, no. 5 (2008): 398–410. https://doi.org/10.1016/j.biotechadv.2008.03.005.

Gutierrez, Maria Paz. "Programming Selective Elasticity." In *Green Design, Materials and Manufacturing Processes*, edited by Helena Bartolo, and Paulo Jorge Da Silva Bartolo, Nuno Manuel Fernandes Alves, and Henrique Amorim Almeida. CRC PressI Llc, 2013.

Gutierrez, Maria-Paz. "Reorienting Innovation: Transdisciplinary Research and Building Technology." *Architectural Research Quarterly* 18, no. 01 (2014): 69–82. https://doi.org/10.1017/S1359135514000372.

Gutierrez, Maria Paz, and Tarek I. Zohdi. "Effective Reflectivity and Heat Generation in Sucrose and PMMA Mixtures." *Energy and Buildings* 71 (March 2014): 95–103. https://doi.org/10.1016/j.enbuild.2013.11.046.

Gutierrez, M. P. "Material Alignments." *Technology|Architecture + Design* 5, no. 1 (2021): 115–18. https://doi.org/10.1080/24751448.2021.1887682.

Jeska, Simone. *Transparent Plastics: Design and Technology*. Springer, 2008. http://books.google.com/books?hl=en&lr=&id=0OML58qoE0IC&oi=fnd&pg=PA7&dq=Transparent+Plastics+Design+and+Technology&ots=GDi8rl7VTA&sig=FWpPZ_56fROeC9hSja5joUGB4aE.

Jones, Mitchell, Andreas Mautner, Stefano Luenco, Alexander Bismarck, and Sabu John. "Engineered Mycelium Composite Construction Materials from Fungal Biorefineries: A Critical Review." *Materials & Design* 187 (February 2020): 108397. https://doi.org/10.1016/j.matdes.2019.108397.

Jung, Patrick, Karen Baumann, Dina Emrich, et al. "Lichens Bite the Dust – A Bioweathering Scenario in the Atacama Desert." *iScience* 23, no. 11 (2020): 101647. https://doi.org/10.1016/j.isci.2020.101647.

Kellersztein, Israel, Daniel Tish, John Pederson, Martin Bechthold, and Chiara Daraio. "Multifunctional Biocomposite Materials from *Chlorella Vulgaris* Microalgae." *Advanced Materials* 37, no. 1 (2025): 2413618. https://doi.org/10.1002/adma.202413618.

Laing, N. "The Use of Solar and Sky Radiation for Air Conditioning of Pneumatic Structures." IASS, *Proceedings of the 1st International Colloquium on Pneumatic Structures,* University of Stuttgart, Stuttgart, Germany, 1967, 11–12.

Lantada, Andrés Díaz, Louisa Estadieu, Michael Gorki, Philipp Höfele, and Monsur Islam. "Blurring the Boundaries between Living and Non-Living? Examining and Mapping the Living Properties of Engineered Matter and Systems." *Technology in Society* 81 (June 2025): 102822. https://doi.org/10.1016/j.techsoc.2025.102822.

Lantada, Andrés Díaz, Jan G. Korvink, and Monsur Islam. "Taxonomy for Engineered Living Materials." *Cell Reports Physical Science* 3, no. 4 (2022): 100807. https://doi.org/10.1016/j.xcrp.2022.100807.

LeCuyer, Annette W., Ian Liddell, Stefan Lehnert, and Ben Morris. *ETFE: Technology and Design*. Birkhäuser, 2008.

Lee, Hsiao Wei, Seyed Ali Rahmaninezhad, Li Meng, et al. "Prediction of Microbial-Induced Calcium Carbonate Precipitation for Self-Healing Cementitious Material." *Cement and Concrete Composites* 158 (April 2025): 105945. https://doi.org/10.1016/j.cemconcomp.2025.105945.

Lewis, W. J. *Tension Structures: Form and Behaviour*. Thomas Telford, 2003.

Lode, Anja, Felix Krujatz, Sophie Brüggemeier, et al. "Green Bioprinting: Fabrication of Photosynthetic Algae-Laden Hydrogel Scaffolds for Biotechnological and Medical Applications." *Engineering in Life Sciences* 15, no. 2 (2015): 177–83. https://doi.org/10.1002/elsc.201400205.

Lorenz, Christian, Elisabetta Bianchi, Giovanni Poggiali, et al. "Survivability of the Lichen Xanthoria Parietina in Simulated Martian Environmental Conditions." *Scientific Reports* 13, no. 1 (2023): 4893. https://doi.org/10.1038/s41598-023-32008-6.

Meikle, Jeffrey L. *American Plastic: A Cultural History*. Rutgers University Press, 1997.

Molinari, Sara, Robert F. Tesoriero, and Caroline M. Ajo-Franklin. "Bottom-up Approaches to Engineered Living Materials: Challenges and Future Directions." *Matter* 4, no. 10 (2021): 3095–3120.

Muggia, Lucia, Theodora Kopun, and Martin Grube. "Effects of Growth Media on the Diversity of Culturable Fungi from Lichens." *Molecules* 22, no. 5 (2017): 824. https://doi.org/10.3390/molecules22050824.

Nazaroff, William W. "Indoor Particle Dynamics." *Indoor Air* 14 no. Supplement 7 (2004): 175–83.

Palmqvist, Kristin, Lena Dahlman, Anna Jonsson, and Thomas H. Nash. "The Carbon Economy of Lichens." In *Lichen Biology, Second Edition*, edited by Thomas H. Nash, III. Cambridge University Press, 2008. https://doi.org/10.1017/CBO9780511790478.011.

Paoli, Luca, Elena Bandoni, and Luigi Sanità Di Toppi. "Lichens and Mosses as Biomonitors of Indoor Pollution." *Biology* 12, no. 9 (2023): 1248. https://doi.org/10.3390/biology12091248.

Park, Younggeun, Maria Paz Gutierrez, and Luke P. Lee. "Reversible Self-Actuated Thermo-Responsive Pore Membrane." *Scientific Reports* 6, no. 1 (2016). https://doi.org/10.1038/srep39402.

Reese, Stefanie. "Meso-Macro Modelling of Fibre-Reinforced Rubber-like Composites Exhibiting Large Elastoplastic Deformation." *International Journal of Solids and Structures* 40, no. 4 (2003): 951–80.

Ritter, Axel. *Smart Materials: In Architecture, Interior Architecture and Design*. Springer, 2007.

Robinson-Gayle, S. "ETFE Foil Cushions in Roofs and Atria." *Construction and Building Materials* 15, no. 7 (2001): 323–27.

Samir, Aya, Fatma H. Ashour, A. A. Abdel Hakim, and Mohamed Bassyouni. "Recent Advances in Biodegradable Polymers for Sustainable Applications." *Npj Materials Degradation* 6, no. 1 (2022): 68. https://doi.org/10.1038/s41529-022-00277-7.

Satchanska, Galina, Slavena Davidova, and Petar D. Petrov. "Natural and Synthetic Polymers for Biomedical and Environmental Applications." *Polymers* 16, no. 8 (2024): 1159. https://doi.org/10.3390/polym16081159.

Schrlau, Jill E., Linda Geiser, Kimberly J. Hageman, Dixon H. Landers, and Staci Massey Simonich. "Comparison of Lichen, Conifer Needles, Passive Air Sampling Devices, and Snowpack as Passive Sampling Media to Measure Semi-Volatile Organic Compounds in Remote Atmospheres." *Environmental Science & Technology* 45, no. 24 (2011): 10354–61. https://doi.org/10.1021/es202418f.

Semper, Gottfried. *The Four Elements of Architecture and Other Writings*. Cambridge University Press, 1989.

Shaeffer, Ronald E. *Tensioned Fabric Structures: A Practical Introduction*. The Society, 1996.

Spribille, Toby, Philipp Resl, Daniel E. Stanton, and Gulnara Tagirdzhanova. "Evolutionary Biology of Lichen Symbioses." *New Phytologist* 234, no. 5 (2022): 1566–82. https://doi.org/10.1111/nph.18048.

Spribille, Toby, Göran Thor, Fred L. Bunnell, Trevor Goward, and Curtis R. Björk. "Lichens on Dead Wood: Species-Substrate Relationships in the Epiphytic Lichen Floras of the Pacific Northwest and Fennoscandia." *Ecography* 31, no. 6 (2008): 741–50. https://doi.org/10.1111/j.1600-0587.2008.05503.x.

Srubar, Wil V. "Engineered Living Materials: Taxonomies and Emerging Trends." *Trends in Biotechnology* 39, no. 6 (2021): 574–83. https://doi.org/10.1016/j.tibtech.2020.10.009.

Steiner, Hadas. "The Forces of Matter." *The Journal of Architecture* 10, no. 1 (2005): 91–109.

Tang, Tzu-Chieh, Bolin An, Yuanyuan Huang, et al. "Materials Design by Synthetic Biology." *Nature Reviews Materials* 6, no. 4 (2021): 332–50. https://doi.org/10.1038/s41578-020-00265-w.

Tripathi, Ankita H., Somya Mehrotra, Amrita Kumari, et al. "Lichens as Bioremediation Agents—A Review." In *Synergistic Approaches for Bioremediation of Environmental Pollutants: Recent Advances and Challenges*, edited by Riti Thapar Kapoor and Maulin P. Shah. Elsevier, 2022. https://doi.org/10.1016/B978-0-323-91860-2.00015-4.

Tully, John A. *The Devil's Milk: A Social History of Rubber*. Monthly Review Press, 2011.

Van Dessel, Steven, and Abdol R. Chini. "Rigidified Pneumatic Composites: Use of Space Technologies to Build the next Generation of American Homes." National Science Foundation Award Abstract #0122022, n.d.

Wang, Yanyi, Yi Liu, Jing Li, Yue Chen, Sizhe Liu, and Chao Zhong. "Engineered Living Materials (ELMs) Design: From Function Allocation to Dynamic Behavior Modulation." *Current Opinion in Chemical Biology* 70 (October 2022): 102188. https://doi.org/10.1016/j.cbpa.2022.102188.

Weiss, Jonathan D., Alana Mermin-Bunnell, Fredrik S. Solberg, et al. "A Low-Cost, Open-Source 3D Printer for Multimaterial and High-Throughput Direct Ink Writing of Soft and Living Materials." *Advanced Materials*, Wiley Online Library, 2025, 2414971.

Wilkinson, Sara J., Peter J. Ralph, and Nimish Biloria. "Algae Building: Is This the New Smart Sustainable Technology?" In *Data-Driven Multivalence in the Built Environment*, edited by Nimish Biloria. S.M.A.R.T. Environments. Springer International Publishing, 2020. https://doi.org/10.1007/978-3-030-12180-8_12.

Wolkoff, Peder. "Indoor Air Humidity, Air Quality, and Health – An Overview." *International Journal of Hygiene and Environmental Health* 221, no. 3 (2018): 376–90. https://doi.org/10.1016/j.ijheh.2018.01.015.

Wolkoff, Peder, Kenichi Azuma, and Paolo Carrer. "Health, Work Performance, and Risk of Infection in Office-like Environments: The Role of Indoor Temperature, Air Humidity, and Ventilation." *International Journal of Hygiene and Environmental Health* 233 (April 2021): 113709. https://doi.org/10.1016/j.ijheh.2021.113709.

Wu, Dong Yang, Sam Meure, and David Solomon. "Self-Healing Polymeric Materials: A Review of Recent Developments." *Progress in Polymer Science* 33, no. 5 (2008): 479–522. www.sciencedirect.com/science/article/pii/S0079670008000208.

Xia, Xiaoxing, Christopher M. Spadaccini, and Julia R. Greer. "Responsive Materials Architected in Space and Time." *Nature Reviews Materials* 7, no. 9 (2022): 683–701.

Zhu, Yingrui, Shan Guo, and Weihui Liang. "A Literature Review Investigating the Impact of Temperature and Humidity on Volatile Organic Compound Emissions from Building Materials." *Building and Environment* 262 (August 2024): 111845. https://doi.org/10.1016/j.buildenv.2024.111845.

PART II
Becoming

Ehsan Baharlou

Becoming explores innovative uses of synthetic polymers and composite materials in architecture, emphasizing sustainability, circularity, adaptability, and advanced fabrication techniques to showcase how designers cautiously address environmental challenges. Becoming illustrates how designers and architects actively participate in the evolution of polymers, from reclamation to bio-based innovation, and redefine design approaches by capturing polymer characteristics in their work.

In *Sustainable Polymers in Architecture: From Recycling to Bio-Based Innovation*, the authors map the evolution of reclaimed synthetic polymers to bio-based polymers. They highlight the integration of digital fabrication technologies with enhanced recycled plastics and bio-based polymers for use in building construction. A series of case studies emphasizes the role of material performance and environmental ethics in transforming plastics in architecture toward a post-carbon era of material circularity and environmental responsibility.

In *Infundibuliforms: Experiments in Spatial Plasticity*, the authors explore spatial plasticity through technological advancements that support the integration of biologically inspired morphologies with advanced thermoplastic elastomers and robotic additive manufacturing. This integration offers real-time experiences of responsive and flexible architectural forms, while challenging the static spatial concept by creating programmable spatial plasticity.

Continuing the exploration of spatial plasticity, the lightness properties of plastic materials are examined in *Hybrid Composite Manufacturing: Plastic Futures for Architecture*, where Marshall Prado explores advanced robotic construction techniques to showcase the lightweight and high performance of fiber-reinforced polymers in architectural structures. This hybrid system, as an innovative approach, challenges conventional construction by integrating material intelligent and design adaptability to reduce material use and situates itself within the circular economy principles.

The design adaptation within a polymer hybrid system is further highlighted by the use of algorithmic design inspired by natural structures. In *Nature's Blueprint: Redefining Plastic Designs*, Julia Koerner thoroughly explores the synergy between nature and plastics. Through innovative, nature-inspired designs, natural patterns are translated into algorithmic design using advanced fabrication techniques, challenging the mass production of plastic with algorithmically customized production.

DOI: 10.4324/9781003613589-7

This part ends with *Plastics Plus*, where Virginia San Fratello explores the combination of bio-based polymers with various waste materials to create sustainable solutions. This innovative approach adds value to plastic production by repurposing wastes from agriculture to wildfires, developing new composite polymers that introduce cultural and sensory values, from familiar scents to familiar textures and colors.

6

SUSTAINABLE POLYMERS IN ARCHITECTURE

From Recycling to Bio-Based Innovation

Philip F. Yuan, Jiapan Sun, and Liming Zhang

Introduction: The Environmental Ethics of Plastic Futures

Entering the 21st century, human society faces growing environmental and resource crises. While globalization has accelerated economic growth, it has also intensified the extraction and depletion of natural resources. Plastics epitomize both industrial progress and ecological burden—once hailed as a revolution in materials science, now a major challenge to sustainability. Since the mid-20th century, their light weight, strength, durability, and low cost have driven widespread use in packaging, construction, transport, and healthcare. Global production exceeds 350 million tons a year, consuming 4–8% of global petroleum output, reflecting deep dependence on synthetic materials.[1]

This prosperity has exacted heavy environmental costs. Most plastics resist degradation for centuries, causing long-term pollution in soil, air, and water. Disposal through landfilling or incineration leads to secondary pollution and greenhouse gas emissions, while recycling remains inefficient due to complex sorting and material decay.[2] In many developing regions, waste is still burned or dumped, worsening ecological risks.[3] The United Nations Environment Programme reports that about eight million tons of plastic enter the oceans annually, threatening marine ecosystems and human health. The true environmental and ethical costs of plastics far exceed their economic value.

Recycled Synthetics: Current Development, Advantages, and Challenges

Under the global drive for sustainability, recyclable plastics have emerged as a key category of green building materials. Their light weight, toughness, corrosion resistance, and formability enable applications across building envelopes, interiors, and urban facilities. Recycled polyethylene (rPE) and polypropylene (rPP) are used in wall panels, roofing membranes, and insulation boards, while modified polyvinyl chloride (rPVC) and polyethylene terephthalate (rPET) appear in flooring and partitions. In landscape and municipal design, recycled plastics are fabricated into benches, planters, and road barriers, realizing the functional reuse of waste. However, limited strength, flame retardancy, and weather resistance—along with heterogeneous

DOI: 10.4324/9781003613589-8

TABLE 6.1 Hierarchical Classification Strategy of 3D-Printed Components Made from Recycled Plastics

Component Type	*Definition*	*Performance Requirements*	*Recycling Cycles*
Structural Components	Used as elements subjected to continuous structural loads	High mechanical strength High weather resistance High fire resistance	0–1
Decorative Components	Used primarily for aesthetic purposes without sustained external loading, such as facade panels or wall claddings	High visual quality High weather resistance High fire resistance Moderate mechanical strength	1–2
Functional Components	Used to perform specific functions, such as molds, pipelines, or acoustic absorption elements	High weather resistance Moderate fire resistance Moderate mechanical strength	1–3
Art Installations and Furniture	Used as temporary artistic installations or non-permanent furniture products	Moderate visual quality Moderate weather resistance Moderate fire resistance Moderate mechanical strength	More than 1

quality—remain key constraints, making performance degradation after multiple recycling cycles a major barrier to broader adoption.

During repeated use, polymer chains undergo scission and oxidation, reducing molecular weight and mechanical performance. Yet, with advances in digital fabrication and 3D printing, such degradation can be reinterpreted as a design opportunity rather than material failure. Our research proposes a graded component system for 3D-printed recycled plastics, allocating materials by residual performance: rPET or rPA for semi-structural parts, modified rPP and rPVC for decorative uses, and rPE or rPS for installations and furniture. This hierarchical utilization extends material lifecycles and exemplifies a circular, design-driven approach to sustainable construction (Table 6.1).

Building upon this foundation, our team developed the MetaPrinter flexible manufacturing system, an integrated 3D printing platform supporting the intelligent application of recycled plastics. Designed for prefabricated architectural components, MetaPrinter provides a full-process workflow that unifies material processing, software control, and equipment configuration. To address the variability of recycled plastics, it incorporates an adaptive process parameter algorithm that dynamically adjusts printing settings—such as layer height, extrusion flow rate, and printhead speed—based on material characteristics and geometric conditions. This enables real-time optimization before and during printing, ensuring stable quality and precision across diverse material types.

In terms of hardware, MetaPrinter adopts a multi-process modular architecture to achieve scalability and process flexibility. The system comprises six standardized modules—five functional and one integrated enclosure—that can be independently replaced or reconfigured according to production needs. Modular interfaces allow efficient switching between different materials, extrusion systems, or robotic setups, facilitating seamless adaptation to various component geometries and fabrication scales. Through this combination of adaptive software and modular hardware, MetaPrinter establishes a robust foundation for sustainable, intelligent, and customizable digital construction workflows (Figure 6.1).

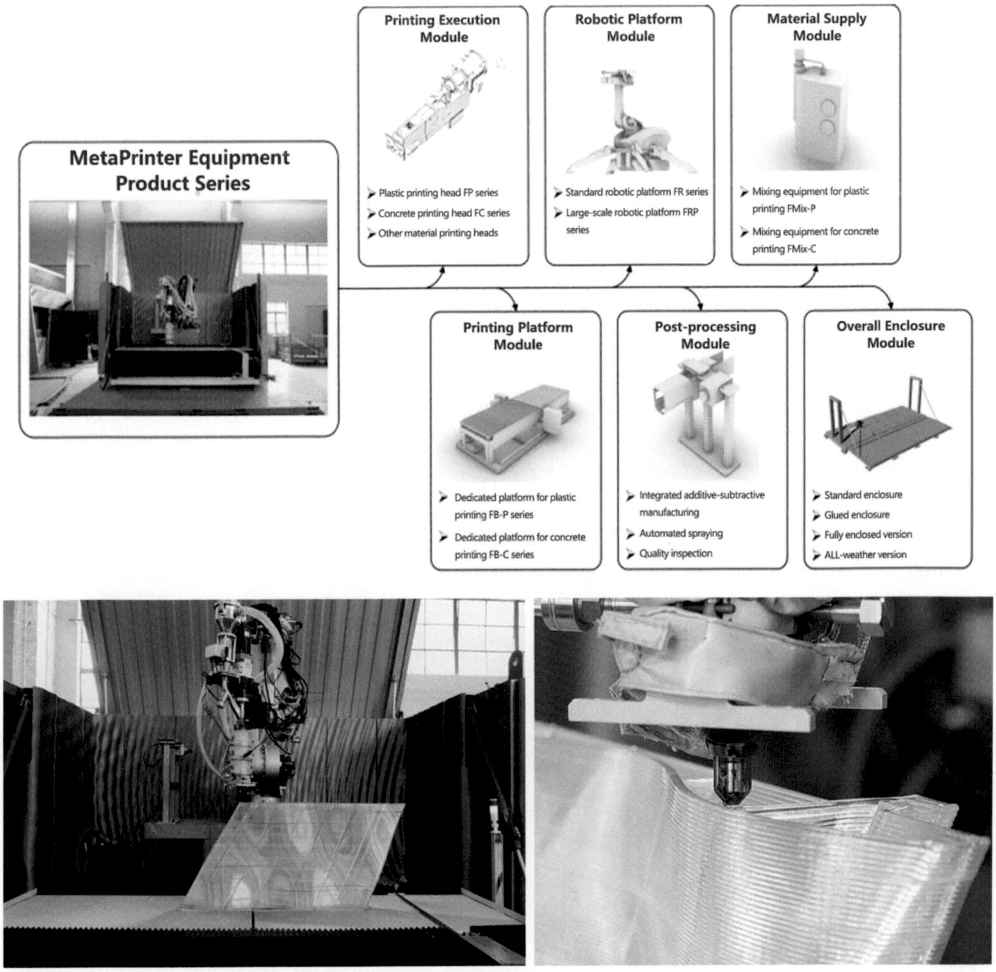

FIGURE 6.1 Modular equipment system and digital twin control software. Illustration by Fab-union.

In terms of control software, MetaPrinter is equipped with a digital twin–based control system. Through the integration of a five-zone pressure stabilization sensor embedded in the extruder, a mass flow meter, and an online optical monitoring unit, the system enables real-time monitoring of the temperature, pressure, mass flow, and surface quality of the plastic melt during the printing process. Coupled with the adaptive process parameter algorithm, this system maintains consistent productivity and printing quality under varying performance conditions of different recycled plastic materials.

Applications of Recycled Synthetics in Architecture

The China Pavilion at the 2018 Venice Architecture Biennale

The "Cloud Village" (Yunshi) project is one of the major practical achievements of the Digital Design and Construction Research Center at Tongji University since 2015 in the fields of

TABLE 6.2 Project Information of the "Cloud Village"

Project Name	*"Cloud Village"—Outdoor Pavilion of the China Pavilion at the 2018 Venice Architecture Biennale*
3D Printing Application Scenario	Integrated structural and decorative components
3D Printing Application Scale	400 m^2
3D Printing Material	Recyclable ABS
Recycling Cycles	1

FIGURE 6.2 Completed view of the "Cloud Village". Photograph by Fab-union.

architectural 3D printing and recycled material research. It also represents a significant milestone in the application of recyclable plastics to structural components in architecture. The project was exhibited in the China Pavilion at the 2018 Venice Architecture Biennale, under the theme "*Our Countryside*." It aimed to explore regenerative pathways for rural architecture in China through digital design and material innovation within the context of the new era.

In this work, the research team systematically demonstrated for the first time in an international exhibition the architectural application of a novel recyclable plastic—modified ABS (acrylonitrile–butadiene–styrene copolymer; Figure 6.2). The material combines excellent structural performance with refined aesthetic quality, serving as a crucial medium for achieving an integration of structural functionality and formal expression in architectural fabrication (Table 6.2).

Compared with the experimental "Cloud Pavilion" prototype, the "Cloud Village" project achieved significant advances in both material performance and component organization. The modified ABS, derived from high-impact copolymer ABS, incorporated a brominated epoxy resin–antimony trioxide flame-retardant system, with hindered amine light stabilizers (HALS) and UV absorbers enhancing weather resistance. Glass fiber and calcium carbonate were introduced for composite reinforcement, markedly improving flexural strength and stiffness while preserving printability. Experimental results showed a tensile strength of 52 MPa, a flexural modulus of 2.6 GPa, and a heat-deflection temperature above 90°C, enabling the material to maintain structural stability under sustained loads. This performance range supports the reliable use of modified ABS in medium-scale architectural components, providing a robust material foundation for the broader structural application of architectural 3D printing (Figure 6.3).

In terms of structural design strategy, the "Cloud Village" project adopted a node displacement optimization method based on the gradient descent algorithm. This optimization approach, developed by the research team of Academician Xie Yimin, is a form-finding method specifically tailored for curved surface network structures. In essence, it enhances the overall stiffness of the surface by iteratively adjusting the positions of structural nodes along their normal vectors—similar to how a corrugated or undulating surface exhibits greater resistance to deformation than a flat one.

Through this algorithm, the surface geometry becomes subtly corrugated after optimization, thereby improving its structural stability. In the "Cloud Village" project, since the roof design includes extensive cantilevered portions, this node displacement optimization was applied to strengthen the roof's deformation resistance. Computational results demonstrated that the maximum nodal displacement of the roof was reduced by approximately 45% after optimization.

At the manufacturing level, the Cloud Village project integrated the FUrobot control platform with the MetaPrinter flexible manufacturing system to achieve non-planar robotic 3D printing with real-time path optimization and feedback control. This workflow enhanced surface continuity and geometric precision while enabling rapid, modular deployment across different fabrication scenarios. Through multi-material extrusion and dynamic temperature regulation, the system maintained stable thermal conditions during ABS printing, improving interlayer adhesion by over 30% and minimizing warping and delamination. The resulting components exhibited high structural integrity, smooth surfaces, and consistent coloration, demonstrating the practical convergence of intelligent control and flexible manufacturing in architectural-scale additive construction.

The architectural system of the Cloud Village pavilion consists of modular structural units that collectively form a lightweight, continuous spatial morphology. Each module is joined through bolted and mortise–tenon-type mechanical connections, enabling rapid on-site assembly and disassembly while exemplifying the reversibility of recycled material systems in circular construction. Components range from 8 to 15 mm in thickness, with individual weights below 40 kg, allowing manual installation without heavy machinery. This lightweight yet high-strength structural strategy reduces construction energy demand and demonstrates the scalability of 3D-printed components for small- to medium-scale architecture, temporary installations, and mobile structures—extending the material and technical logic of sustainable digital fabrication beyond the experimental pavilion context.

The "Cloud Village" project verified the feasibility of using newly developed recyclable plastics in structural components and marked a pivotal shift in architectural 3D printing from experimental research to engineering-scale application. The successful demonstration showed that modified recycled plastics can meet both structural and aesthetic requirements, redefining plastics as materials capable of uniting functionality and design articulation. Beyond material validation, the project established a closed-loop framework integrating material, design, and fabrication, exemplifying a new paradigm for synergy between digital construction and circular materials. Through this process, the team proposed an intelligent recycling system for architectural applications—combining performance grading, algorithmic optimization, and equipment coordination—to enable structured reuse of recycled plastics. Extending this model to facades, pavilions, and urban infrastructure suggests a broader vision in which sustainable

FIGURE 6.3 Material testing and joint prototyping for the "Cloud Village" project. Photograph by Fab-union.

transformation depends not only on inventing new materials but on reprogramming existing ones through digital manufacturing and design innovation.

East Gate of Nanjing Happy Valley Theme Park, China

The construction of the East Gate of Nanjing Happy Valley Theme Park represents the first large-scale application of recyclable ABS material in a real architectural environment following the experimental projects “Cloud Pavilion” and “Cloud Village.” Unlike the previous explorations that primarily focused on exhibition and installation contexts, the East Gate serves as the main entrance of the theme park, functioning simultaneously as a node for pedestrian convergence and dispersion, a spatial interface, and an urban symbol, while also being required to meet long-term outdoor durability and weather resistance standards.

This project demonstrates that recyclable ABS, when properly modified and optimized through advanced processing techniques, can effectively support the fabrication of large and geometrically complex structures, showcasing significant advantages in real engineering applications (Table 6.3 and Figure 6.4).

At the design level, the East Gate serves both as a functional architectural structure and as an urban spatial interface. Its form was required to maintain iconic visual identity while fulfilling practical functions such as visitor circulation, shading, and wayfinding. To balance functionality, structure, and form, the project adopted a layered construction system consisting of a primary steel framework, secondary support subframes, and outer ABS 3D-printed panels.

The steel structure provides overall load-bearing capacity and stability, while the secondary framework accommodates construction tolerances and facilitates geometric transitions. The outer ABS panels articulate the complex curved surfaces of the facade. This stratified system ensures structural reliability while enabling formal flexibility, thereby fully demonstrating the plasticity and expressive potential of 3D-printed polymer materials in architectural applications (Figure 6.5).

During construction, one of the key challenges was fabricating large-scale curved components within the constraints of robotic arm reach and printing scale. The team adopted a modular strategy, decomposing the overall geometry into manageable units assembled from multiple ABS 3D-printed panels. Fabrication was conducted in a controlled factory environment, followed by on-site assembly. A hierarchical error control system ensured precision across different levels: meter-scale for the primary structure, centimeter-scale for the secondary, and millimeter-scale for the ABS panels. The material’s flexibility allowed minor on-site adjustments, enabling accurate alignment and adaptive assembly, which departed from the conventional “design-then-assemble” model and demonstrated the responsive logic of digital construction.

TABLE 6.3 Project Information of the East Gate of Nanjing Happy Valley Theme Park

Project Name	*East Gate of Nanjing Happy Valley Theme Park, China*
3D Printing Application Scenario	Decorative components
3D Printing Application Scale	1,950 m^2
3D Printing Material	Recyclable ABS
Recycling Cycles	1

FIGURE 6.4 Completed view of the East Gate of Nanjing Happy Valley Theme Park. Photograph by Fab-union.

To ensure long-term outdoor performance, the research focused on enhancing the weather resistance and durability of ABS. A modified formulation containing UV stabilizers, antioxidants, and heat stabilizers significantly reduced yellowing and embrittlement under hot-humid conditions. Process optimization further improved interlayer adhesion and surface density, increasing water resistance and mechanical reliability. Accelerated aging tests of 3,360 hours confirmed that the modified ABS retained over 80% of its original strength, validating its feasibility for architectural facades in Nanjing's climate.

In terms of form and aesthetic expression, the lightweight and highly formable properties of ABS provided favorable conditions for the efficient realization of complex curved surfaces. Unlike concrete or metal cladding panels, ABS panels require no costly molds and can be fabricated directly from digital models, thereby simplifying the design-to-construction workflow and reducing manufacturing costs. The material's low weight also facilitates transportation and on-site installation, enabling more flexible and efficient assembly of intricate geometries. In addition, by applying a color-gradient algorithm to the panel surfaces, the design team achieved a visual balance between overall unity and localized variation, reinforcing the symbolic and recognizable character of the architectural entrance.

East Gate of Nanjing Happy Valley Theme Park project holds significant practical importance as it marks a key transition of 3D-printed recyclable ABS from experimental research to real architectural application. Through strategies such as modular decomposition, hierarchical error control, and flexible assembly, the project provides a new methodological framework for digital

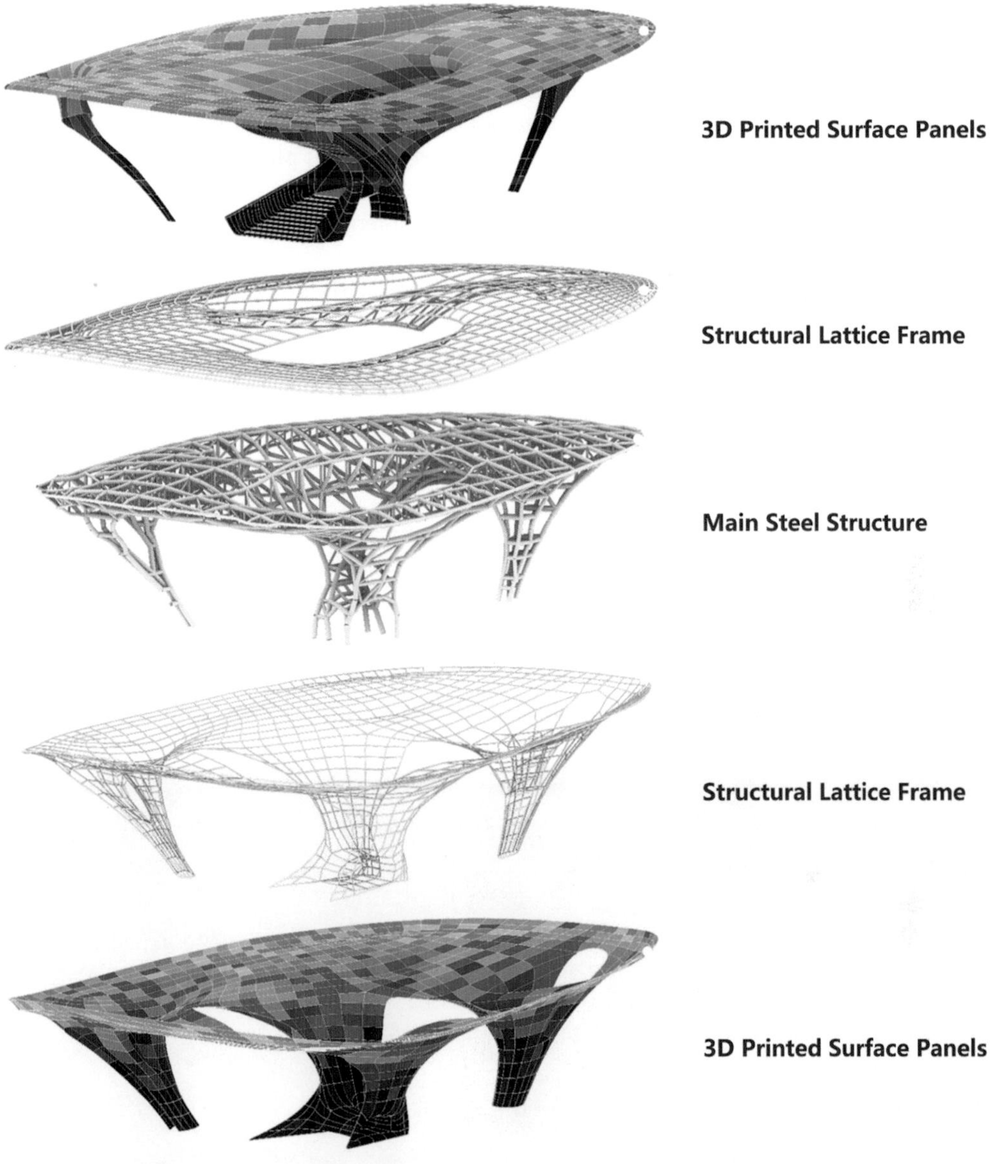

FIGURE 6.5 Exploded view of the layered construction system of the East Gate of Nanjing Happy Valley Theme Park. Illustration by Fab-union.

construction of large-scale and geometrically complex structures. Meanwhile, its advantages in recyclability, lightweight strength, construction efficiency, and improved weather resistance establish ABS as an important medium linking green architecture and digital fabrication. This case not only expands the application boundaries of polymer materials in architecture but also offers a technical reference and practical paradigm for the industrialization and sustainable development of future buildings centered on recyclable materials.

Red Pavilion at the Light of Internet Expo Center, China

In the "Cloud Village" and "Happy Valley" projects, recyclable plastics were applied in architectural contexts as structural and decorative components, respectively, demonstrating the high-performance potential of recycled materials in architectural 3D printing. However, with an increasing number of recycling cycles, the polymer chains of plastics undergo progressive scission and degradation, leading to a significant decline in mechanical strength, surface quality, and weathering stability, making them unsuitable for long-term use or multifunctional structural applications.

Against this backdrop, the research team shifted its focus toward the functional applications of multi-recycled plastics within the construction process, exploring their potential roles in formwork systems and temporary structural supports. The "Red Pavilion" at the Light of Internet Expo Center in Wuzhen stands as a representative practice developed under this research direction (Table 6.4 and Figure 6.6).

TABLE 6.4 Project Information of the Red Pavilion at the Light of Internet Expo Center

Project Name	*Red Pavilion at the Light of Internet Expo Center, China*
3D Printing Application Scenario	Functional component: formwork
3D Printing Application Scale	330 m^2
3D Printing Material	Recyclable ABS
Recycling Cycles	2

FIGURE 6.6 Completed view of the Red Pavilion at the Light of Internet Expo Center. Photograph by Fab-union.

The core objective of the "Red Pavilion" project was to achieve precise and controllable fabrication of freeform surfaces within a large-span thin-shell structural system, while ensuring cost efficiency and sustainability throughout the construction process. Since the final load-bearing capacity of the shell structure is provided by a brick–polyurethane composite system, the 3D-printed material itself was not required to serve as a permanent structural component.

Therefore, the research team selected a hybrid mixture of multi-recycled ABS and rPP as the 3D-printed formwork material. Although this material no longer meets architectural standards in terms of strength, toughness, and weather resistance, it still possesses good flowability, moderate thermal stability, and high formability, making it suitable for high-precision shaping. By remelting and extruding waste plastic pellets, the team not only realized material reuse but also established a traceable circular material pathway, enabling the recycled plastic to serve a dual role as both forming medium and functional component during the construction phase (Figure 6.7).

During the geometric design phase, the research team adopted a dual form-finding strategy combining graphic statics and a particle–spring algorithm to generate a compression-dominant shell geometry. Using integrated RhinoVAULT–Kangaroo computation, the form was optimized for equilibrium and discretized into about 1,500 modular units. Each module's base surface was planarized to ensure printability while allowing controlled discontinuities to maintain overall geometric flexibility. This workflow ensured both precision and structural stability, providing a reliable computational foundation for large-span thin-shell construction.

From a fabrication perspective, the concept of "formwork as component" was introduced, enabling 3D-printed formwork to serve as both casting medium and part of the final structure. The formwork, made of multi-recycled ABS/rPP reinforced with short glass fibers, was coated with a cast polyurethane insulation layer and robotically milled to achieve the designed double-curved geometry. Modular units were factory-fabricated by robotic extrusion, casting, and milling with a tolerance of ±2 mm, then assembled on site through bolted joints. The printed system collaborated with an inward-offset steel frame, maintaining a final deviation within ±20 mm, demonstrating a precise and efficient workflow for sustainable construction.

To verify the structural safety of the construction, the research team conducted an on-site load test after completion. Approximately 90 tons of sandbags were placed over the walkable area of the shell to monitor the structural response and deformation under different loading conditions. The test results indicated that both the formwork system and the overall structure remained stable throughout the loading process, with no visible cracking or local buckling observed (Figure 6.8). These findings demonstrate that the 3D-printed formwork possesses sufficient load-bearing capacity and geometric precision for use in large-scale thin-shell construction.

The completion of the "Red Pavilion" project illustrates a viable downcycling reuse pathway for multi-recycled plastics in architectural 3D printing. Although these materials no longer possess the mechanical strength or long-term durability required for structural applications, their processability and controllability allow them to serve effectively during the construction phase. By transforming performance-degraded plastics into functional elements such as formwork, insulation layers, and temporary supports, the project extended material lifecycles and achieved near closed-loop utilization, reducing both virgin material consumption and construction waste. This research demonstrates that even plastics approaching their physical limits can acquire renewed functional value within digital fabrication systems, offering a practical strategy for low-carbon, circular development, and providing technical guidance for the sustainable use of recycled plastics in architectural fabrication.

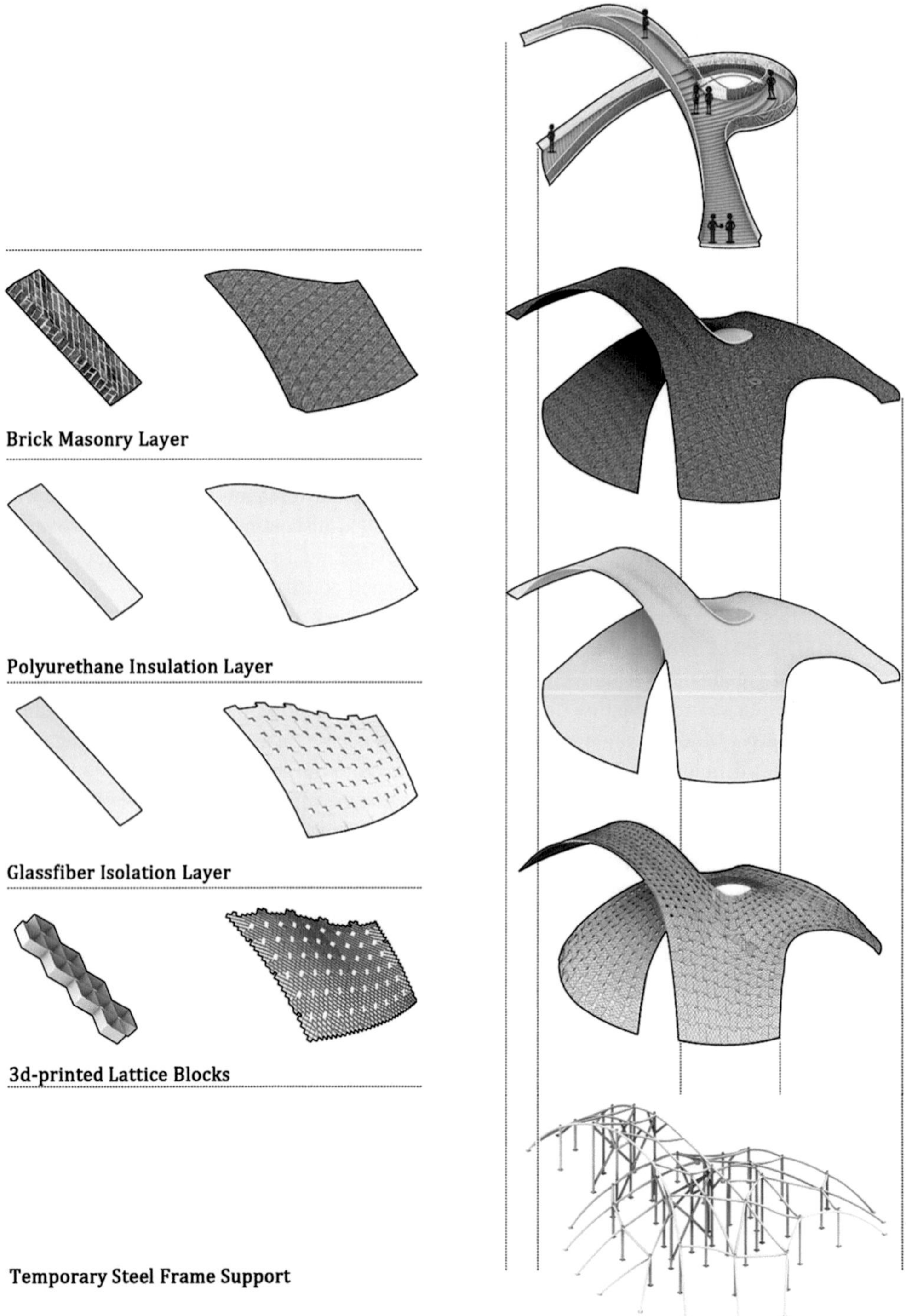

FIGURE 6.7 Exploded view of the layered construction system of the Red Pavilion at the Light of Internet Expo Center. Illustration by Fab-union.

FIGURE 6.8 On-site construction of the "Red Pavilion" at the Light of Internet Expo Center, Wuzhen. Photograph by Fab-union.

Outdoor Facilities at the NIKE Office Building, China

Following a series of large-scale architectural projects that validated the engineering potential of recyclable plastics in both structural and functional components, the research team began exploring the reapplication of such materials in small-scale urban contexts. In 2023, a collaborative project with NIKE was completed, featuring a series of public art installations and urban furniture, marking an important step in extending the use of recyclable plastics in architectural 3D printing from structural experimentation to everyday urban life.

This series utilized discarded sports shoes as the primary raw material, and through large-scale fused deposition 3D printing technology, achieved a fully regenerative workflow—from material recovery and form generation to component fabrication. The works demonstrate the potential of plastics at the end of their lifecycle to re-enter urban construction in new and meaningful forms (Table 6.5 and Figure 6.9).

In terms of material sourcing, the plastics used in the project were derived from recycled footwear waste collected through NIKE's "Grind" program. These discarded materials typically consist of a mixture of rubber, thermoplastic polyurethane (TPU), and polyethylene (PE), among other polymers. After undergoing processes of crushing, separation, and thermal regranulation, the recovered fragments were remanufactured into extrudable filaments suitable for large-scale 3D printing.

Due to the complex composition and limited purity of the source materials, their mechanical strength and thermal stability could not match those of single-polymer systems. Consequently, the project did not employ them as structural materials but instead defined them as non-load-bearing materials for urban furniture and artistic installations. This "end-of-life reutilization" strategy allowed plastics that are otherwise difficult to recycle to gain new functionality through digital fabrication, thereby extending their social and material value.

The project integrates recycled material properties with parametric form generation, using algorithmic control of density, texture, and toolpath to balance form and performance. Made from multi-recycled NIKE "Grind" plastics, the design includes double-layer sculptures and corrugated benches optimized through topological deformation and motion data. Fabricated via the MetaPrinter and FUrobot platforms, the process achieved adaptive printing at 220°C with over 95% material efficiency and minimal waste. Installed in public spaces, the lightweight components combine ergonomic precision and expressive form while promoting environmental awareness. The work demonstrates how digital fabrication transforms end-of-life plastics into functional, durable, and creative urban artifacts, providing a model for sustainable, low-carbon design.

Towards Bio-Based Polymers

The Rise, Properties, and Challenges of Bio-Based Polymers

From a recycling perspective, the "sustainability" of plastics does not imply infinite reusability. Each physical recycling cycle causes polymer chain scission and oxidation, reducing molecular weight and mechanical strength, eventually making materials unsuitable for structural use.[4] Additives such as pigments and plasticizers further complicate sorting and remelting, limiting both technical and economic feasibility.[5] Even after repeated recycling, plastics degrade slowly

TABLE 6.5 Project Information of Outdoor Facilities at the NIKE Office Building

Project Name	*Outdoor Facilities at the NIKE Office Building, China*
3D Printing Application Scenario	Application in small-scale art installations and urban furniture
3D Printing Material	Recyclable ABS
Recycling Cycles	3

FIGURE 6.9 Completed view of NIKE's "Grind" program. Photograph by Fab-union.

in the natural environment, persisting for decades or centuries.[6] This exposes the paradox of recycling—caught between the ideal of circularity and the reality of material decay.

Bio-fabrication, by contrast, offers a forward-looking paradigm that generates materials through biological synthesis rather than reconstructing waste. Microbial fermentation and cultivation techniques enable the production of bio-based polymers such as polylactic acid (PLA) and polyhydroxyalkanoates (PHAs). PLA, derived from starch or agricultural residues, is easily processed but brittle and heat-sensitive, while PHAs, metabolically produced by microorganisms, are fully biodegradable in soil and marine environments.[7] Though costly and limited in scale, both represent materials designed for renewal and degradation, reducing fossil dependence and microplastic pollution.[8] For architectural use, PLA suits decorative or temporary components, while PHAs are promising for environmentally adaptive applications. Yet challenges remain—low heat and fire resistance, limited mechanical durability, and high cost—rendering their large-scale architectural adoption still experimental but emblematic of a new biocentric material vision.

Applications of Bio-Based Polymers in Architecture

Ruixue Multi-Hall and the Taihu Rock in the Master-of-Nets Garden, Suzhou

Although PLA exhibits limited aging and thermal resistance—becoming brittle and deformable under UV exposure or fluctuating humidity—its environmental advantages have inspired further sustainable strategies. Recognizing the inherent durability limitations of bio-based polymers, our team explored the integration of natural fibers into recyclable plastics as an alternative green solution. Incorporating materials such as wood flour or plant fibers not only enhances toughness, heat resistance, and dimensional stability but also reduces the reliance on pure petroleum-based polymers. Combined with stabilization treatments—including UV absorbers, hindered amine light stabilizers (HALS), and antioxidants—this hybrid approach improves overall weatherability and mechanical integrity. It demonstrates how coupling recycled polymers with bio-based reinforcements can create durable yet sustainable materials for architectural applications (Figure 6.10).

In the Ruixue Multi-Hall project, modified ABS was reinforced with plant fibers, reducing polymer chain scission and pigment degradation under UV and humidity. The results confirmed that fiber-modified ABS meets the durability and aesthetic requirements of indoor and semi-open environments, offering practical validation for bio-based plastics in facade systems. The 3D-Printed Taihu Rock in Suzhou's Master-of-Nets Garden further tested PLA under extreme conditions. With hydrophobic agents and toughening additives, the modified PLA achieved greater water resistance, flexibility, and dimensional stability during prolonged immersion, retaining tensile strength and resisting hydrolysis. Serving as both a sculptural and material experiment, the project demonstrated PLA's adaptability in landscape construction and cultural contexts, illustrating how bio-based polymers can merge technological precision with environmental and artistic expression (Figure 6.11).

Overall, these two practices provided empirical evidence for the use of PLA in architectural contexts. While PLA remains unsuitable for large-scale structural or long-term facade applications, targeted modifications have endowed it with the potential to perform

FIGURE 6.10 Ruixue Multi-Hall in Tianfu Agricultural Expo Park, China. Photograph by Archi-Union.

FIGURE 6.11 The Taihu Rock in the Master-of-Nets Garden, Suzhou. Photograph by Fab-Union.

in small- to medium-scale landscape structures and temporary installations. These cases point toward the broader role of bio-based materials in sustainable architectural pathways of the future.

The Biohybrid Pavilion at DigitalFUTURES 2025: Modified Applications of PHA Materials

In 2025, the College of Architecture and Urban Planning at Tongji University realized the Biohybrid Pavilion at the DigitalFUTURES International Workshop, integrating biodegradable materials with plant growth systems (Figure 6.12). The structure used modified polyhydroxyalkanoate (PHA) bioplastics, 3D-printed into three 2.5-meter hollow columns serving as the primary supports. To overcome PHA's weak strength and low weather resistance, reinforcing fillers and stabilizers were added to enhance load-bearing performance. Planting grooves on the columns enabled Parthenocissus tricuspidata (Boston ivy) to gradually envelop the pavilion, transferring structural dependence from bioplastics to living vegetation as degradation progressed. This experiment, the first large-scale application of modified PHA in architecture, proposed a cyclic model of "disappearing structures and growing supports," redefining material decay as ecological succession and illustrating a new paradigm of synergy between digital fabrication and ecological processes.

Design Concepts for Using Bio-Based Polymers as Riverbank Components

Bio-based polymers represent a potential material for 3D-printed ecological riverbank structures, offering a novel concept for future waterfront restoration. In this vision, bio-based polymers are processed through large-scale 3D printing into modular riverbank units with porosity, steps, and textured surfaces. These features not only satisfy the basic requirements of structural stability but also create conditions for aquatic vegetation to root and for microorganisms to colonize. Over time, the polymer components would gradually degrade, enabling the riverbank to achieve a form of "self-renewal" through the alternation of artificial and natural systems: in the early stage, the printed structures maintain form and stability, while in the later stage, plant root systems and sediment deposits gradually assume the structural role. This design reconceptualizes riverbanks not as static protective infrastructure but as evolving ecological interfaces, simultaneously addressing protective needs and fostering ecological restoration and symbiosis within aquatic environments.

Conclusion

In summary, the evolution from recycled to bio-based polymers reflects a gradual shift in architectural material logic—from mitigating waste to regenerating ecosystems. Recycled polymers, despite their limitations, provide a realistic transitional route that bridges current industry practices with sustainable goals. Bio-based polymers, though still experimental, embody the next phase of this transformation, promising materials that are both renewable and degradable. As digital fabrication and material innovation converge, the future of architectural materials lies not only in replacing petrochemical plastics but in reimagining construction as a process of ecological cohabitation.

FIGURE 6.12 The Biohybrid Pavilion at DigitalFUTURES 2025. Photograph by DigitalFUTURES.

Notes

1 Laura Da Pires Mata Costa et al., "Capture and Reuse of Carbon Dioxide (CO_2) For a Plastics Circular Economy: A Review," *Processes* 9, no. 5 (2021), https://doi.org/10.3390/pr9050759.
2 Ahmed Z. Naser et al., "Poly(Lactic Acid) (PLA) And Polyhydroxyalkanoates (PHAs), Green Alternatives to Petroleum-Based Plastics: A Review," *RSC advances* 11, no. 28 (2021), https://doi.org/10.1039/D1RA02390J.
3 Kalpana Gopinath et al., "Quantification of Microplastic in Red Hills Lake of Chennai City, Tamil Nadu, India," *Environmental science and pollution research international* 27, no. 26 (2020), https://doi.org/10.1007/s11356-020-09622-2.
4 Adolfo Le Pera et al., "Effect of Quality of Separately Collected Glass, Paper Plus Cardboard and Lightweight Packaging Waste on Environmental, Energetic and Economic Sustainability of the Material Recovery Facility Operations," *Journal of Cleaner Production* 425 (2023), https://doi.org/10.1016/j.jclepro.2023.138973.
5 John N. Hahladakis et al., "An Overview of Chemical Additives Present in Plastics: Migration, Release, Fate and Environmental Impact During Their Use, Disposal and Recycling," *Journal of hazardous materials* 344 (2018), https://doi.org/10.1016/j.jhazmat.2017.10.014.
6 A. L. Andrady, "Persistence of Plastic Litter in the Oceans," in Marine Anthropogenic Litter, eds. M. Bergmann, L. Gutow, and M. Klages (Cham: Springer, 2015), 57–72.
7 Sameh S. Ali et al., "Plastic Wastes Biodegradation: Mechanisms, Challenges and Future Prospects," *Science of the total environment* 780 (2021), https://doi.org/10.1016/j.scitotenv.2021.146590.
8 Yolanda González Garcia et al., "Síntesis Y Biodegradación De Polihidroxialcanoatos: Plásticos De Origen Microbiano," *Revista internacional de contaminación ambiental* 29, no. 1 (2013).

References

Ali, Sameh S., Tamer Elsamahy, and Rania Al-Tohamy et al. "Plastic Wastes Biodegradation: Mechanisms, Challenges and Future Prospects." *Science of the Total Environment* 780 (2021): 146590. https://doi.org/10.1016/j.scitotenv.2021.146590.

Beaumont, Nicola J., Margrethe Aanesen, and Melanie C. Austen et al. "Global Ecological, Social and Economic Impacts of Marine Plastic." *Marine Pollution Bulletin* 142 (2019): 189–95. https://doi.org/10.1016/j.marpolbul.2019.03.022.

Da Pires Mata Costa, Laura, Débora Micheline Vaz de Miranda, and Ana Carolina Couto de Oliveira et al. "Capture and Reuse of Carbon Dioxide (CO_2) For a Plastics Circular Economy: A Review." *Processes* 9, no. 5 (2021): 759. https://doi.org/10.3390/pr9050759.

González Garcia, Yolanda, Juan Carlos Meza Conteras, Orfil González Reynoso, and Jesús Antonio Córdova López. "Síntesis Y Biodegradación De Polihidroxialcanoatos: Plásticos De Origen Microbiano." *Revista internacional de contaminación ambiental* 29, no. 1 (2013): 77–115.

Gopinath, Kalpana, Srinivasalu Seshachalam, and K. Neelavannan et al. "Quantification of Microplastic in Red Hills Lake of Chennai City, Tamil Nadu, India." *Environmental Science and Pollution Research International* 27, no. 26 (2020): 33297–306. https://doi.org/10.1007/s11356-020-09622-2.

Hahladakis, John N., Costas A. Velis, Roland Weber, Eleni Iacovidou, and Phil Purnell. "An Overview of Chemical Additives Present in Plastics: Migration, Release, Fate and Environmental Impact During Their Use, Disposal and Recycling." *Journal of Hazardous Materials* 344 (2018): 179–99. https://doi.org/10.1016/j.jhazmat.2017.10.014.

Le Pera, Adolfo, Miriam Sellaro, Giulia Grande, Egidio Bencivenni, and Massimo Migliori. "Effect of Quality of Separately Collected Glass, Paper Plus Cardboard and Lightweight Packaging Waste on Environmental, Energetic and Economic Sustainability of the Material Recovery Facility Operations." *Journal of Cleaner Production* 425 (2023): 138973. https://doi.org/10.1016/j.jclepro.2023.138973.

Naser, Ahmed Z., I. Deiab, and Basil M. Darras. "Poly(Lactic Acid) (PLA) and Polyhydroxyalkanoates (PHAs), Green Alternatives to Petroleum-Based Plastics: A Review." *RSC Advances* 11, no. 28 (2021): 17151–96. https://doi.org/10.1039/D1RA02390J

7

INFUNDIBULIFORMS

Experiments in Spatial Plasticity

Kathy Velikov, Wes McGee, Daniel Tish, and Geoffrey Thün

Introduction

We are compelled by architectures that redefine relations between humans, matter, space, and the environment. While architecture is traditionally a practice concerned with static structures, we are interested in exploring the potentials of architectures whose boundaries resist a singular stable spatial enclosure and instead pursue spatial metastability, where spatial conditions can never be fixed. We might term this a kind of *spatial plasticity* where not only space, but potentially the immaterial spatial matters such as light, atmosphere, and acoustics, are continually shaped and reshaped by fluctuating boundary conditions. In biology, form is never stable. Ernst Haeckel, the German biologist and naturalist who coined the term 'ecology' and was one of the first to document the morphological transformation of organisms as a co-evolutionary process in relation to their environment, noted that, "Nothing is constant but change! All existence is a perpetual flux of being and becoming".[1] The compelling lessons of biology for architecture, then, extend to more than the intelligence of biological structures and the properties of biogenic materials, but also to ideas of metastability, plasticity, and states of becoming. The *Infundibuliforms* project experiments with producing a spatially plastic enclosure that dynamically transforms in its spatial form and volume. To achieve this, the project also advances plastic material systems: specifically elastomeric cable nets (Figure 7.1).

Lightweight and adaptive structures, such as cable nets, also have a history in biological principles. Some of the earliest advances were undertaken at the University of Stuttgart's Institute for Lightweight Structures where, under the directorship of engineer Frei Otto, began to methodically study and experiment with the structural principles of biological organisms for buildings toward increased material efficiency and performative capabilities.[2] The lightweight structures advanced by Otto, such as minimal nets and pneus, were dependent on the physics of form-active structures and the material properties of textiles and membranes. Advancements in the design of tensile and membrane structures were facilitated by the evolution of synthetic polymers for architectural applications, which surged in the period between the 1930s and 1970s, and enabled the realization of lightweight buildings, large-scale pneumatic structures, and transformable building components.[3]

DOI: 10.4324/9781003613589-9

FIGURE 7.1 Video stills of various positions of the *Infundibuliforms* kinetic tensile installation. Photos by the authors.

Lightweight and adaptable architectures, such as those incorporating membranes and inflatables, have not only been a technical concern in the field but have also been mobilized to expand the expectations and conceptions of building. This includes not only the idea that a building might resist the Vitruvian notion of "firmitas" but also in terms of agonistic positions that challenge architecture's status quo.[4] Activist architects and artists of the 1960s embraced the ephemeral bubble as oppositional to the rigid political and social climate of the times, and inflatables became associated both with technological and social utopias.[5] In their seminal book on Kinetic Architecture, William Zuk and Roger Clark state that "our present task is to unfreeze architecture, to make it a fluid, vibrating, changeable backdrop for the various and constantly changing modes of life. An expanding, contracting, pulsating, changing architecture would reflect life as it is today and therefore be a part of it".[6] Our work in responsive building envelopes has been often motivated by the dual interests in advancing technologies for adaptive and responsive architectural skins toward architectural performance for environmental adaptation while simultaneously exploring the relational, aesthetic, and emotive capabilities of fluctuating and dynamic material systems.[7]

Carrying these investigations forward, *Infundibuliforms* is a kinetic cable-net installation capable of large-scale deformation into variable configurations. "Infundibulum" is a biological term that refers to funnel-shaped cavities or organs, such as the petals of morning glory flowers or the sucker cavities of octopus limbs. The capacity of these biological structures to physically deform in real time for a variety of purposes was both an instrumental and conceptual basis for this project to develop a spatially-scaled, kinetic surface. The full-scale demonstrator installation consisted of a 3D-printed elastomeric tensile net whose height and inclination could be actively varied, with three infundibula that could each be dynamically reconfigured between vaulted and chimney forms. The research aims of the project were to explore the material, formal, and operational possibilities of active tensile surfaces. Our disciplinary aims were to expand on the technical and spatial possibilities of an architecture whose main virtue lies in its ability to be in-firm; to embrace change as an ally toward a spatial plasticity that "dynamically responds to performative and programmatic vicissitudes with nuanced geometries and continuous gradients of form".[8]

Project Development

The project was developed through an iterative process spanning research domains of kinetic actuation, robotic additive manufacturing, material exploration in thermoplastic elastomers (TPE), topology design, material programming, and the cable robot control environment. Through these explorations, the project serves as a case study for technological advancements in variable spatial plasticity, as well as material research in thermoplastic manufacturing of functionally graded cable nets.

Kinetic Actuation

The method of actuation is one of the primary considerations for a kinetic structure as the nature and capacity of the actuator is a determinant for the transformation of the system. While localized, small-scale actuators such as pneumatic actuators or servo and stepper motors are able to produce modest variations in surface form and texture, or changes in degree of form, this project aimed to produce more dramatic spatial changes in form within the topology of the cable net. After exploring multiple actuation options, we opted to advance a cable robot system that could be leveraged to produce the larger-scale deformation the project sought to achieve.[9] The installed demonstrator consisted of an ovoid steel outer ring that tensioned the 28m^2 cable-net surface, with three weighted steel inner rings attached to the end of each infundibulum. Three sets of cables with customized winches were attached to each ring, with their lengths controlled by individual servo motors. With this setup, three-dimensional positioning of the control rings in space is possible. While the positioning envelope of each control ring is constrained by the location of the servo motors in plan, this system offers a wide range of vertical displacements if given enough cable length and mounting height (Figure 7.2).

As the cable robot system afforded such large displacements, the challenge quickly became the design of a surface that could negotiate the different states offered by the actuation system. In this installation, the surface area could change by over 50%, and this became a key driver for the design. Elasticity, or the ability for a material to regain its original shape after stretching, and the amount a material can stretch before breaking, were central properties to accommodate the changing surface conditions while minimizing slack and sag during transitions. These properties are almost exclusively found in rubbers and elastomeric plastics and drove the fabrication research toward investigations of soft, stretchy plastics and manipulations that could expand upon and demonstrate their innate qualities as an envelope typology. By tapping into the ability of TPE material to change its length and then recover its original form after stretching, we could design with flexibility for this application (Figure 7.3).

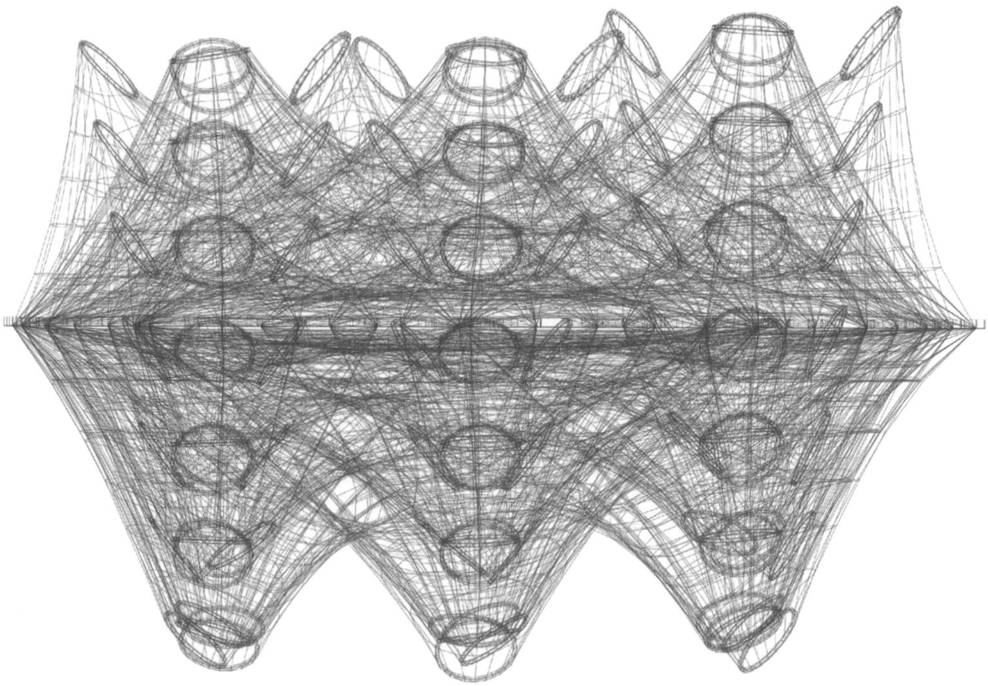

FIGURE 7.2 Model wireframe overlay of position extents of the three infundibula. Drawing by the authors.

FIGURE 7.3 Detail of cable robot controlled inner ring at one infundibulum, with another seen in the distance. Photo by the authors.

Robotic Additive Manufacturing

Since the first tents and fishing nets, tensile surfaces have most typically been composed of knit or woven fabrics, membranes, or mechanically fastened nets. Additive manufacturing (i.e. robotic 3D printing) offers an alternative means of manufacture for tensile surfaces and, at the time this project was undertaken, had been relatively underexplored for cable nets.[10] Additive manufacturing enables the advancement of developing net structures with programmable material behaviors and opens the possibility of novel performative geometries.[11] However, conventional additive manufacturing approaches, such as Fused Filament Fabrication (FFF), are not well-suited to extremely soft materials as they require a minimum stiffness to drive the filament through the heat chamber and extruder nozzle, otherwise risking the buckling of the filament and jamming of the mechanism. Therefore, a custom pellet-based extruder was designed and built to facilitate this work. Pellet extruders can additively manufacture a broader range of materials as the pellets are gravity-fed into the extrusion chamber, where they are melted for deposition. They also boast substantially higher material deposition rates, allowing larger components to be produced in less time.

Our method deposits the liquefied TPE mesh pattern onto a heated aluminum bed. The heated bed keeps the TPE viscous enough to facilitate joint fusion at the intersections while the surface is being printed, and then allows for controlled cooling into a monolithic textile surface (Figure 7.4).

FIGURE 7.4 Deposition of a TPE mesh panel onto a heated surface with the robotic extruder. Photo by the authors.

Material Exploration

Leveraging the broad material palette available with pellet extrusion, a range of soft, elastic polymers were tested for their applicability in our cable-net structure. TPE are a broad category of stretchy, thermoplastic polymers used in a wide range of industries, from toothbrushes to automotive door seals. Within the broader category of TPEs, block copolymers are materials made from two or more monomers (monomers being the building blocks of polymers), notable for their inherent toughness and their tunable elasticity. These properties are highly valued in this application to allow the tensile net to be stretched to greater degrees without risk of failure. By working with material suppliers capable of custom compounding, we tested multiple TPE block copolymers in a range of durometers, a measure of softness or hardness, and melt flow indices, a measure of extrudability. A high melt flow index was found to assist with the extrusion process, allowing material to flow faster and more consistently than lower melt flows. Two durometers of block copolymers with high melt flow indices were then selected for the installation: a very soft 35A durometer and a stiffer 68A durometer block copolymer material. These two grades of polymer are selectively applied across the installation's surface according to shifts in performance requirements across the continuous surface. The softer material is designated for the infundibular sections of the surface that undergo the most dramatic change in surface area, utilizing the material's ease of deformation without building too much stress within the surface. The stiffer material is then applied through the midsection of the piece to hold tension within the surface and prevent excessive self-deformation under gravity.

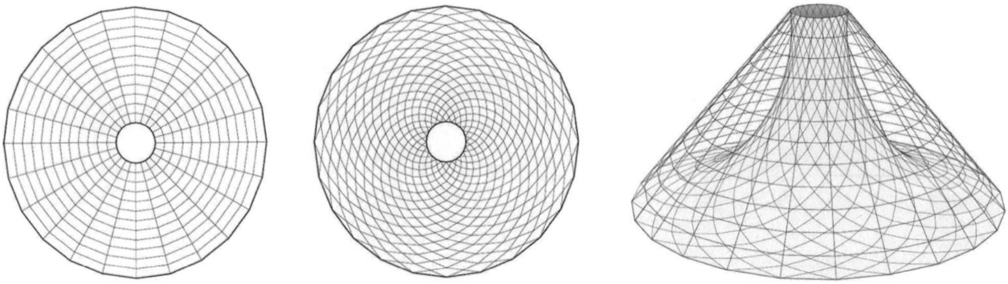

FIGURE 7.5 Form-active shapes overlay (r) of quad mesh (l) and diagonal grid (m), with identical physics parameters applied to the cables by a particle-spring simulation system. Drawing by the authors.

Topology Design

The geometric challenge of this project was to create an elastic cable net geometry that could be extruded onto a flat surface and then stretched into specifically tailored anticlastic (negative curvature at all points) forms of the infundibula. As a form active structure, the design of the cable net topology, or the connective structure of the net edges, has a direct impact on the flow of forces and, thus, the possible range of forms of the surface. When working with radial/concentric mesh topologies, the resulting form-active surface is distinctly conical due to the direct load path from the control ring to the outer boundary. This mesh topology also leads to inefficient material utilization, as the radial elements take the vast majority of the tension loads while the concentric elements receive very little loading.

Introducing a diagrid in place of the conventional radial grid structure is advantageous in many ways. First, the diagrid leads to a more uniform material utilization as both directions of the diagrid carry loads from the control ring to the outer boundary. Additionally, the diagrid produces a mesh topology with exclusively even-valenced vertices, eliminating the need for start-and-stop extrusion procedures. In contrast, the standard radial topology produces an odd number of line segments at the outer edges of the mesh. Finally, the spiraling nature of the load paths through the diagrid creates the desired anticlastic form. This active form may even be tuned to create more or less acutely anticlastic shapes by altering the ratio of mesh faces along the U and V axes (Figure 7.5).

More mesh faces in the concentric direction will produce a more dramatically anticlastic form with a tighter funnel. In contrast, a higher ratio of mesh faces in the radial dimension will tend toward conical shapes. This behavior is attributed to the increasingly helical nature of the load path with more relative concentric faces and a more direct load path from the control ring to the outer boundary in the radial dimension. Through this method, the active form of the three funnel-shaped infundibula in the installation was shaped and crafted, introducing two subdivisions in the mesh topology to maintain an appropriate density near the outer boundary of the surface (Figure 7.6).

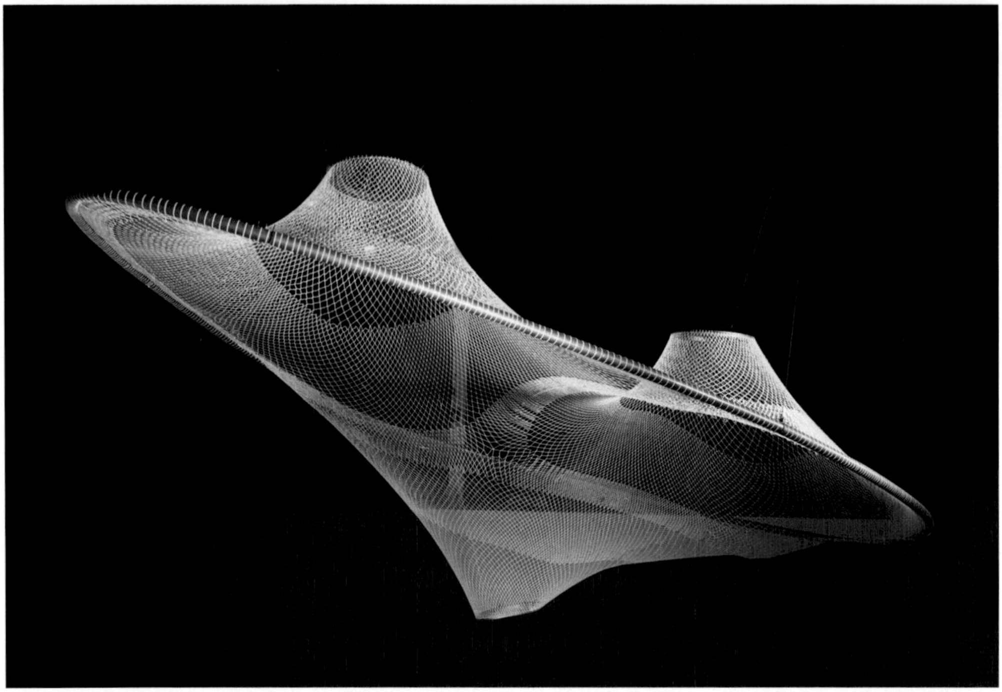

FIGURE 7.6 Mesh density and topology allow for tailored anticlastic forms. Photo by the authors.

Material Programming: Flat2Form

As our additive manufacturing process required printing onto a heated surface to maintain joint strength, all of the panels for the installation would need to be produced in two dimensions only. A common approach to creating three-dimensional forms from two-dimensional materials is the sartorial method of stitching together curved seams to produce out-of-plane forms. This method deals poorly with doubly-curved surfaces, like the project's anticlastic funicular forms, and does not maintain tension as the surface reaches its minimal area when the control rings pass through the plane of the outer boundary. To address these shortcomings, we developed a smart geometry approach to program the 3D form into the printed 2D cable net, which we call *Flat2Form*.

The Flat2Form method is tied directly to insights from the material behavior itself. As the TPE material is soft and flexible, a printed curve can be straightened into a line with minimal force. The material's elasticity allows the curve to be recaptured when the force is removed. Therefore, we can program additional extensibility through geometric means with minimal loading before stretching the material itself with greater force, all working with the unique capacities of the plastic material. By precisely controlling the length of the curved elements that make up the mesh, 3D forms can be programmed into the 2D meshes (Figure 7.7).

The method begins by projecting the 3D form directly into the 2D plane representing the installation plan. The length of every edge in the cable net structure in both 3D and 2D is recorded. The goal of the computational algorithm is to introduce a sinusoidal curve that is the same length as the 3D edge between the endpoints defined by the 2D line. This goal is achieved through

FIGURE 7.7 3D printing TPE of a panel section, showing the curvature of the edges that achieve the Flat2Form geometry. Photo by the authors.

a particle-spring physics system in which each 2D line is subdivided into many equal-length segments connected by free-rotating pin joints that are then iteratively grown (or increased in length) until the sum of the line segments' length matches the length of the original 3D line. The physics engine plays a critical part in this process, allowing the entire mesh to adjust fluidly to avoid unintended intersections that would disrupt the mesh topology. The physics engine includes collision-avoiding radii tuned to the width of the print bead at every point along the line to prevent these intersections. Additionally, the joint angle at each crossing in the mesh topology is preserved and programmed into the mesh through the physics engine to maintain the fused nature of the intersection joints. While these intersection angles are locked, the position of the intersections on the printed plane is not, allowing the mesh vertices to move and self-adjust to accommodate the dramatic increase in material placed near the control rings where the difference between the 3D and 2D forms is greatest. This self-adjustment has no impact on mesh topology and, therefore, no impact on the form-active behavior (Figure 7.8).

As each edge is individually controlled in the algorithm, we can program doubly curved funicular forms into the mesh. This ability to affect how the structure responds to stresses was leveraged in the final *Infundibuliforms* installation. Due to the surface's continuous nature, the cable net's outer sections tend to build up stresses fastest and would naturally produce a slight outward tilt in the control rings. We were able to program additional material into these outward-facing regions of the cable net to allow for extra geometric deformation, neutralizing the tilt of the control rings. This method is uniquely tied to both the capabilities of the material itself and the precision of the robotic additive manufacturing fabrication process, eliciting a novel way to produce doubly-curved funicular forms in 2D production and smart material behaviors at the intersection of material and process (Figure 7.9).

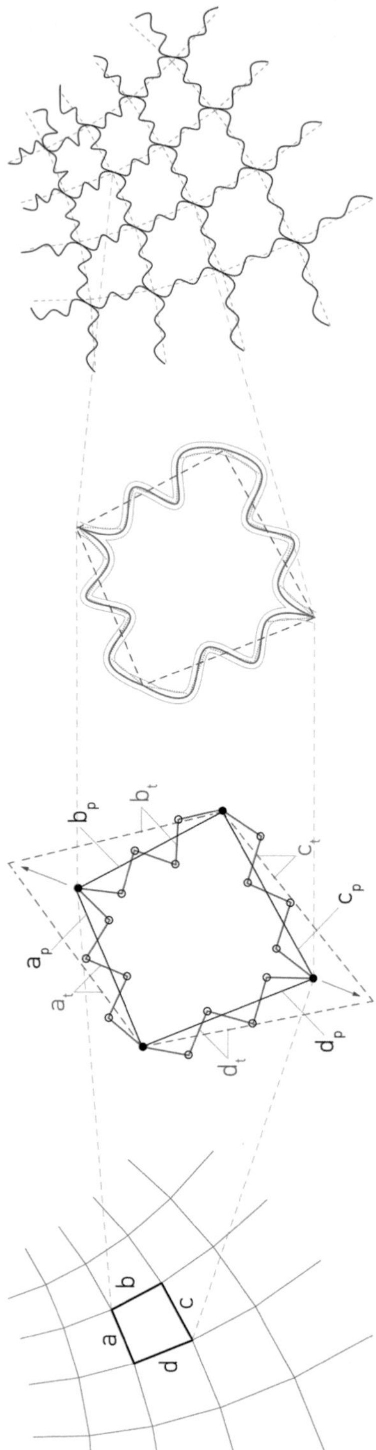

FIGURE 7.8 Process of geometric development for programmable geometry of mesh cell edges. Drawing by the authors.

FIGURE 7.9 Detail of the extended mesh showing how the programmed edge geometry and mesh density produce the specific form. Photo by the authors.

Robotic Control Environment

The simulation system used to predict the form-active shape of the tensile net was expanded upon to become the real-time control system for the live installation. The installation features three cable robots, each consisting of three servo motors connected to a rigid ring and three additional servo motors that can independently raise, lower, and orient the outer steel boundary. In the computational control environment, particle-spring physics simulation allows users to drive and visualize each control ring in three-dimensional space. This input provides a target point that the control rings attempt to reach. However, given the influence of the tensile net, many control ring positions within the theoretical boundary are unattainable. The physics simulation predicts the achievable location of any target point, and the resulting lengths of each cable in the robot are used as the drive lengths for the physical installation. This simulation-in-the-loop motion control paradigm requires fast processing times for the simulation engine but adds an extra layer of safety to ensure that all physical locations are achievable and all cables remain in tension.

The physical installation is controlled by TwinCAT, an industrial Programmable Logic Controller (PLC) and motion control software. A custom plugin written for the Rhinoceros 3D software using the Grasshopper visual programming environment[12] streams the target cable lengths through an open API to the motion controller, which executes the servo positioning loops in real time (Figure 7.10).

Material Agency

Infundibuliforms foregrounds a reorientation of architectural design thinking where matter is not understood as a passive substrate shaped by external design intentions but an active participant in the generation of form, performance, and experience. The project demonstrates how

FIGURE 7.10 Scenes of interaction with various positions of the kinetic installation. Photos by Peter Smith Photography.

programmable, elastic materials—when coupled with computational design, robotic fabrication, and kinetic control logics—can enable architectures that are not only responsive to but co-constitutive with their environments. This reframing of *material as agent* challenges long-held assumptions of fixity and authorship in architectural production. Rather than imposing form upon matter, the design process becomes a negotiation *with* it, allowing form to emerge and be co-produced through material behavior and, in this case, its intrinsic capacities for deformation, memory, and recovery. Plasticity here is not only morphological but ontological: a condition of architecture's capacity to adapt and reconfigure in response to shifting forces.

By embracing the uncertainties and flexibilities of soft, form-active systems, *Infundibuliforms* offers an alternative trajectory for architectures that privileges responsiveness over rigidity, transformation over permanence, and collaboration with material systems over deterministic controls. In doing so, it invites designers to reconceive materiality not as a constraint to be overcome, but as a dynamic, generative force to design *with*—one capable of producing nuanced spatial experiences and adaptive environments attuned to the complexities of contemporary life.

Acknowledgements

Previous versions of portions of the text have been published in Tish et al[13]. OpenAI Chat GPT-4 was used to improve the language of the abstract and conclusion. Funding: Taubman College of Architecture and Urban Planning: 2016 Research Through Making Program, University of Michigan Office of Research: Small Projects Grant. Fabrication/Research Assistants: Asa Peller, Dustin Brugmann, Andrew Kremers, Andrew Wald, Iram Moreno Pinon, Isabelle Leysens. Technical Partners: Buckeye Polymers, Industrial Fabricating Systems, Beckhoff.

Notes

1 Ernst Haeckel, *The Wonders of Life: A Popular Study of Biological Philosophy*, trans. Joseph McCabe (New York: Harper & brothers, 1905).

2 See the series of publications from the Institute for Lightweight Structures (IL), particularly IL3 Biology and Building Part 1 (1971), IL4 Biology and Building Part 2 (1971), IL5 Convertible Roofs (1972), IL6 Biology and Building: Pneus (1974), and IL 25 Form Force Mass: Experiments (1990).

3 Jan Knippers et al., *Construction Manual for Polymers +membranes*: *Materials, Semi-Finished Products, Form-Finding, Design,* Edition Detail (Birkhèauser Verlag, 2011).

4 Reyner Banham, "Monumental Windbags," *New Society* 18 April 1968 (1968).

5 Marc Dessauce, *The Inflatable Moment: Pnuematics and Protest in'68* (New York: Princeton Architectural Press, 1999).

6 William Zuk and Roger H. Clark, *Kinetic Architecture* (New York: Van Nostrand Reinhold, 1970), 9.

7 Geoffrey Thün et al., "The Agency of Responsive Envelopes: Interaction, Politics and Interconnected Systems," *International Journal of Architectural Computing* 10, no. 3 (2012), https://doi.org/10.1260/1478-0771.10.3.377.

Kathy Velikov et al., "Computational and Physical Modeling for Multi-Cellular Pneumatic Envelope Assemblies," *International Journal of Architectural Computing* 13, no. 2 (2015), https://doi.org/10.1260/1478-0771.13.2.143.

Kathy Velikov, "Bio-Physical Envelopes / Sensing Boundaries / Emotive Intermediaries: Material Ecologies of Discursive Skins," in *Bio/matter/Techno/Synthetics: Design Futures for the More Than Human*, ed. Franca Trubiano et al. (New York: Actar Publishers, 2024).

8 Wes McGee et al., "Prologue for a Robotic Architecture: Bridging Design Environments, Robotic Additive Manufacturing, and Systemic Controls for Kinetic Information," in *Towards a Robotic Architecture*, ed. Mahesh Daas and Andrew J. Wit (New York: ORO Editions, 2018), 191.

9 McGee et al., "Prologue for a Robotic Architecture: Bridging design environments, robotic additive manufacturing, and systemic controls for kinetic information," 193.

10 Daniel Tish et al., "Case Studies in Topological Design and Optimization of Additively Manufactured Cable-Nets," *Structures* 18 (2019), https://doi.org/10.1016/j.istruc.2018.11.002.
11 Fergal B. Coulter and Anton Ianakiev, "4D Printing Inflatable Silicone Structures," *3D Printing and Additive Manufacturing* 2, no. 3 (2015).
12 Rhinoceros 3D (often referred to as Rhino) is a CAD software developed by Robert McNeel and Associates. Grasshopper is a visual programming language that runs within Rhino.
13 Tish et al., "Case Studies in Topological Design and Optimization of Additively Manufactured Cable-nets."

McGee et al., "Prologue for a Robotic Architecture: Bridging design environments, robotic additive manufacturing, and systemic controls for kinetic information."

Wes McGee et al., "Infundibuliforms: Kinetic Systems, Additive Manufacturing for Cable Nets, and Tensile Surface Control," in *Fabricate 2017: Rethinking Design and Construction*, ed. Achim Menges et al. (London: UCL Press, 2017).

References

Banham, Reyner. "Monumental Windbags." *New Society* 18 April 1968 (1968): 569–70.

Coulter, Fergal B., and Anton Ianakiev. "4D Printing Inflatable Silicone Structures." *3D Printing and Additive Manufacturing* 2, no. 3 (2015): 140–44.

Dessauce, Marc. *The Inflatable Moment: Pnuematics and Protest in '68*. New York: Princeton Architectural Press, 1999.

Haeckel, Ernst. *The Wonders of Life: A Popular Study of Biological Philosophy*. Translated by Joseph McCabe. New York: Harper & brothers, 1905.

Knippers, Jan, Jan Cremers, Markus Gabler, and Julian Lienhard. *Construction Manual for Polymers +membranes: Materials, Semi-Finished Products, Form-Finding, Design.* Edition Detail. Switzerland: Birkhèauser Verlag, 2011.

McGee, Wes, Kathy Velikov, Geoffrey Thün, and Daniel Tish. "Infundibuliforms: Kinetic Systems, Additive Manufacturing for Cable Nets, and Tensile Surface Control." In *Fabricate 2017: Rethinking Design and Construction*. Edited by Achim Menges et al., 84–91. London: UCL Press, 2017.

McGee, Wes, Kathy Velikov, Geoffrey Thün, and Daniel Tish. "Prologue for a Robotic Architecture: Bridging Design Environments, Robotic Additive Manufacturing, and Systemic Controls for Kinetic Information." In *Towards a Robotic Architecture*. Edited by Mahesh Daas and Andrew J. Wit, 190–203. New York: ORO Editions, 2018.

Thün, Geoffrey, Kathy Velikov, Mary O'Malley, and Lisa Sauvé. "The Agency of Responsive Envelopes: Interaction, Politics and Interconnected Systems." *International Journal of Architectural Computing* 10, no. 3 (2012): 377–400. https://doi.org/10.1260/1478-0771.10.3.377.

Tish, Daniel, Wes McGee, Tim Schork, Geoffrey Thün, and Kathy Velikov. "Case Studies in Topological Design and Optimization of Additively Manufactured Cable-Nets." *Structures* 18 (2019): 83–90. https://doi.org/10.1016/j.istruc.2018.11.002.

Velikov, Kathy. "Bio-Physical Envelopes / Sensing Boundaries / Emotive Intermediaries: Material Ecologies of Discursive Skins." In *Bio/matter/Techno/Synthetics: Design Futures for the More Than Human*. Edited by Franca Trubiano et al., 277–92. New York: Actar Publishers, 2024.

Velikov, Kathy, Geoffrey Thün, Mary O'Malley, and Lars Junghans. "Computational and Physical Modeling for Multi-Cellular Pneumatic Envelope Assemblies." *International Journal of Architectural Computing* 13, no. 2 (2015): 143–68. https://doi.org/10.1260/1478-0771.13.2.143.

Zuk, William, and Roger H. Clark. *Kinetic Architecture*. New York: Van Nostrand Reinhold, 1970.

8

HYBRID COMPOSITE MANUFACTURING

Plastic Futures for Architecture

Marshall Prado

Introduction

Architecture relies heavily on standard material forms like steel, concrete, and wood. While these materials have defined the built environment for centuries, they impose rigid constraints on design and construction. Their inherent inflexibility, standardized forms, and inefficiencies have limited architectural innovation. Concrete, for example, is cast into superfluous molds that dictate its shape. Its environmental issues have been widely documented.[1] Steel beams are produced in standard sizes that prioritize structural efficiency over design freedom. Both concrete and steel have high embodied carbon and energy.[2] Despite its natural versatility and variation, wood is often reduced to dimensional lumber that disregards its anisotropic properties. Such approaches create material systems that resist adaptation and innovation, prioritizing uniformity, and ease of production over performance and customization.

This reliance on static material forms has stifled architectural innovation, particularly as environmental challenges demand more adaptive and sustainable solutions. In his book *Making: Anthropology, Archaeology, Art and Architecture*, Tim Ingold describes the exploration of diverse materials and adaptive design processes in his experiments that reframe the connection between materials, form, and making. He goes on to describe a process of making that aligns with a more morphogenic "process of growth" rather than imposing form on materials.[3] Traditional construction methods often impose form on material rather than leveraging the material's inherent properties. Materials in conventional construction are often treated as passive, dictated by industrial processes. This paradigm contrasts sharply with the potential of fiber-reinforced polymers (FRP), which combine plastics and fibers into composite systems. During fabrication, these materials are typically built-up fiber-by-fiber and layer-by-layer to tailor the composite to meet diverse performance criteria. Unlike traditional systems, FRPs allow for greater geometric freedom, enabling designers and fabricators to craft complex, adaptive forms that respond to specific architectural needs (see Figure 8.1). Furthermore, flexibility, strength, material density, and direction can be controlled, making them ideal for lightweight, high-performance applications. Composite systems embrace material agency, using advanced fabrication techniques to amplify the performance and adaptability of the materials themselves.

DOI: 10.4324/9781003613589-10

FIGURE 8.1 Hybrid composite demonstrator – UTK Filament Tower.

This shift calls for a re-evaluation of architectural production, which moves beyond static materials' limitations to embrace dynamic, integrated systems that align material properties with design intent.

By critiquing the rigidity of traditional construction practices and celebrating the adaptability of composite systems, this article makes a case for plastics and composites as transformative materials for architecture. These systems challenge existing norms and offer a vision of an architectural future driven by innovation, sustainability, and material intelligence (Figure 8.2).

From Experimentation to Industrialization

Plastic's early adoption in architecture was marked by experimentation. In the mid-20th century, plastics were seen as a novel material with limitless possibilities. Designers like Buckminster Fuller explored plastics for their lightweight and moldable properties, creating domes and enclosures that redefined structural possibilities. FRPs played a key role in developing high-performance engineering applications such as those in the aerospace and automotive industries. These materials were celebrated for their exceptional strength-to-weight ratio, anisotropic behavior, and the ability to tailor their performance by controlling the layering of fibers and resins. This layered assembly enables the creation of highly customized structures that can meet precise performance requirements for rigidity and flexibility. These qualities have made fiber composites indispensable for applications where strength, lightness, durability, and efficiency are

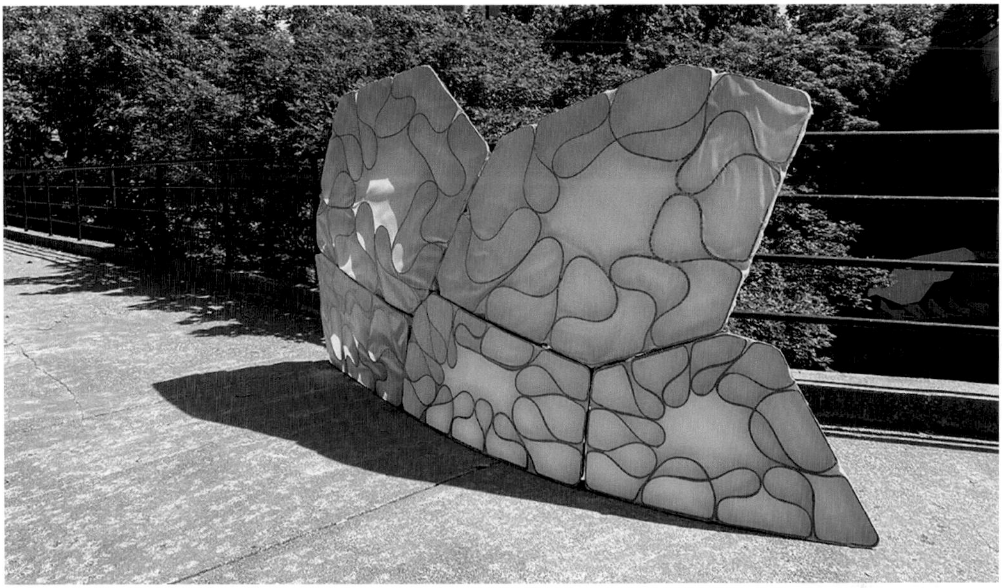

FIGURE 8.2 Hybrid thermoplastic composite building components.

paramount, such as aircraft fuselages, car chassis, and high-performance sporting equipment. In architecture, experimental projects like the Monsanto House of the Future further showcased the potential of composites, demonstrating how they could revolutionize construction by enabling seamless, futuristic forms and lightweight, material-efficient construction.[4] Composites were celebrated for their capacity to break free from the limitations of traditional materials and create novel architectural expressions, offering the promise of a dynamic and innovative design language (Figure 8.3).[5]

However, the novelty of plastics soon faded in public opinion, as the role shifted from experimental material to commodity. The drive for efficiency and profit in a capitalist economy led to their mass production in standardized forms. The advent of injection molding and extrusion processes prioritized speed and cost over innovation and performance. This commodification reduced plastics to a symbol of disposability, with widespread adoption in single-use consumer products reinforcing their reputation as cheap, environmentally harmful materials. Nevertheless, advancements in recycling technologies and bio-based polymers demonstrate how these materials can fit with a circular economy more efficiently and with fewer greenhouse gas emissions than concrete or steel.[6]

The devaluing of plastics as an architectural material undercuts their potential for adaptability and performance. Rather than being limited to cheap, standardized forms, plastics can be considered a key component of adaptive systems that integrate flexibility, strength, and customization. Architects can unlock plastics' capacity to create dynamic, responsive structures that align with modern design and environmental priorities by focusing on the material's potential for customization and hybridization.

Fiber composites have yet to gain significant traction in architecture despite their advantages. This neglect is particularly egregious given that architecture, with its bespoke and context-sensitive nature, stands to benefit significantly from adaptive materials and fabrication processes.

FIGURE 8.3 Monsanto House of the Future at Disneyland.

Unfortunately, traditional composite fabrication relies heavily on molds, which are often prohibitively expensive or materially wasteful for one-off applications. This approach is more feasible in fields like automotive and aerospace, where large production runs justify the cost of tooling. However, the reliance on serialized production for most composites limits the geometric freedom and performance potential. This generates significant waste and unnecessary expenses, making traditional composite production less viable in architectural contexts. The performative opportunities of fiber composites remain untapped due to these production challenges. Fiber composites have struggled to break free from the constraints of serialized manufacturing. This highlights the need for alternative fabrication techniques that align with the one-off nature of architectural production (Figure 8.4).

Emerging technologies, such as robotic fabrication, additive manufacturing (AM), and coreless filament winding,[7] present solutions to these challenges. Robotic fabrication enables precise control of material deposition, for fast, efficient manufacturing. This creates opportunities for adaptive production processes and customized tooling. Additive manufacturing is primarily used for small-scale prototyping, but when combined with industrial robotic arms, it can be used to produce full-scale customizable building components. Coreless filament winding is an innovative robotic process that sequentially builds composite structures fiber by fiber without the need for a large-scale surface mold or winding mandrel. This is possible by winding a continuous fiber across a minimal frame. Iteratively adding ordered fibers in controlled directions can create a tension-based production process for composite manufacturing. This process was originally pioneered at the Institute for Computational Design and Construction at the University of

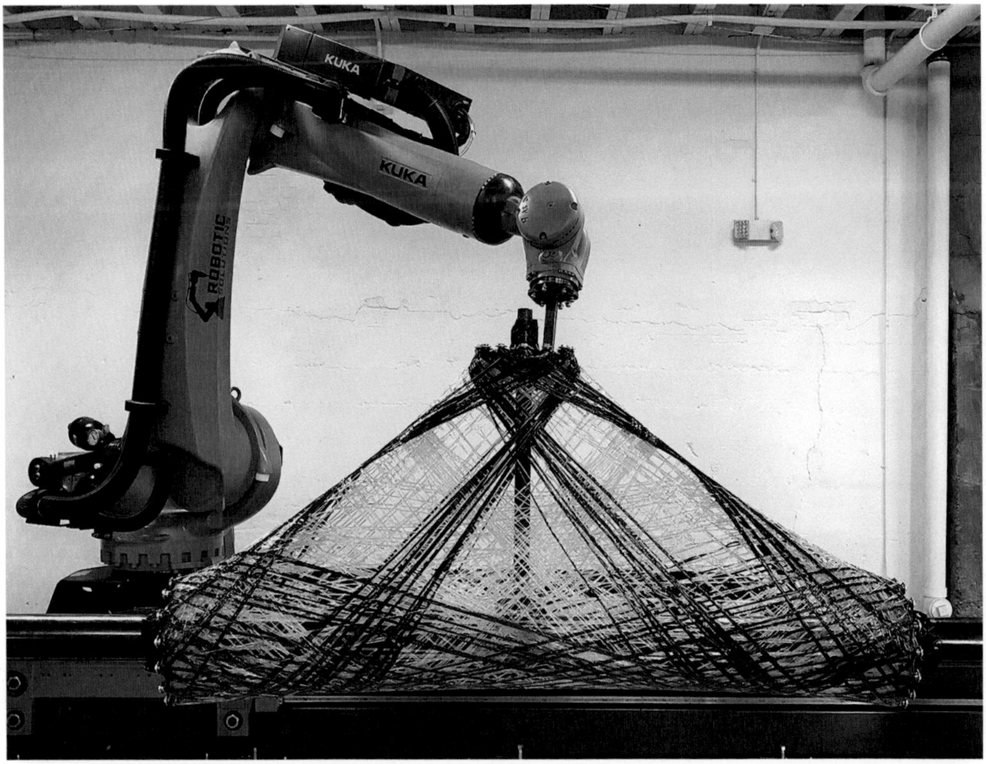

FIGURE 8.4 Coreless filament winding.

Stuttgart and used for several large-scale composite demonstrators.[8] By building polymer and composite structures layer by layer, architects can precisely control not only the form but also the material density, orientation, color, transparency, strength, and rigidity of a design. This granular level of control enables the creation of performative structures tailored to specific environmental and structural needs. Drawing inspiration from biological systems, structural components can effectively be "grown," as Ingold might suggest, integrating all required properties to meet the demands of an architectural application without excess material or waste. These technologies enable the fabrication of complex geometries without the need of traditional molds, reducing material waste and enabling greater design flexibility. These advancements set the stage for a shift in how fiber composites and polymers are utilized in architecture, offering a path toward sustainable, performative, and adaptable construction methods.

These digital technologies do have limitations at an architectural scale. Many of the polymer materials typically used for AM processes are not suitable for large-scale architectural application due to low tensile strength, high flexibility, or low melting temperatures. Although AM processes offer increased geometric freedom over traditional construction processes, they often require significant material investment to achieve strength without fiber reinforcement. This can lead to unnecessarily heavy or inefficient parts at an architectural scale. In the building and construction industries, AM technologies primarily rely on cementitious materials to achieve structural performance. These AM materials add significant weight to AM construction. Furthermore, most architectural spaces have spanning structures, overhangs,

FIGURE 8.5 Large format additive manufacturing.

and/or cantilevered forms, which remain difficult to produce with concrete AM without additional support structures such as precast lintels, shoring, or corbeled geometries. Coreless filament winding, on the other hand, excels in material efficiency but still relies on custom frame structures to define winding geometry. This can limit the geometric freedom of the production process to hyperbolic surfaces, which may not apply to all architectural applications (Figure 8.5).

Combining and hybridizing processes for AM and composite production offer solutions to these limitations while also addressing the performative architectural constraints mentioned earlier. Coreless wound composite structures can be used where structural efficiency is required, while AM is suitable to produce components where the form is most important. A hybrid approach can be used where form and strength are necessary. AM can minimally define a form and save on material resources when combined with composites for strength. Alternatively, composites can span unsupported during fabrication through tension to provide scaffolding for a secondary AM process. These synergistic processes create new opportunities for the fabrication of performative structures using plastics and composites. Hybridizing polymer AM and composite production redefines architectural fabrication, enabling innovative, sustainable, and efficient construction methods. These systems offer a vision of architecture that integrates material intelligence, design flexibility, and performance-driven solutions, paving the way for future advancements in adaptive and hybrid materials.

Hybrid Composite Manufacturing for Architectural Applications

Two research case studies are presented to show the potential of AM and FRPs for architectural applications and the strategies for hybrid composite manufacturing. The UTK Filament Tower is a demonstrator produced for an architectural exhibition in Columbus, Indiana in 2019. It showcases the use of AM and coreless filament winding to produce a large-scale architectural tower structure. Hybrid Thermoplastic-Composite Components were developed to demonstrate novel hybrid composite manufacturing processes for lightweight and adaptive building components (Figure 8.6).

The UTK Filament Tower was based on the biomimetic investigation of a Cholla cactus skeleton. This natural role model provides performative principles that can be adapted to composite production. The cactus structure is comprised of a biaxial lattice that forms a tube-like form. The lattice has a continuous fiber structure layered in various directions to form smooth nodes and intersections. Variations in material density and fiber direction can be observed throughout the structure. These characteristics can be abstracted and transferred to the design and construction logics of a composite tower and share a similar structural topology and organization. This required a fabrication process that could combine material in controlled sequences. A robotic coreless filament winding process was developed to create this cactus-inspired material system at an architectural scale. This process used glass and carbon fibers with an epoxy resin to represent the structure of the cactus fibers. These technical fibers share similar properties to natural fiber composites found in the cactus. The biaxial lattice geometry was split into components to simplify the fabrication process and to fit within the limited working space of an industrial robotic arm. Multi-node component typologies were extracted from the lattice by dividing the struts at the midpoints. Each component has three or four joints that are connected to neighboring components. This multi-node strategy distinguished itself from

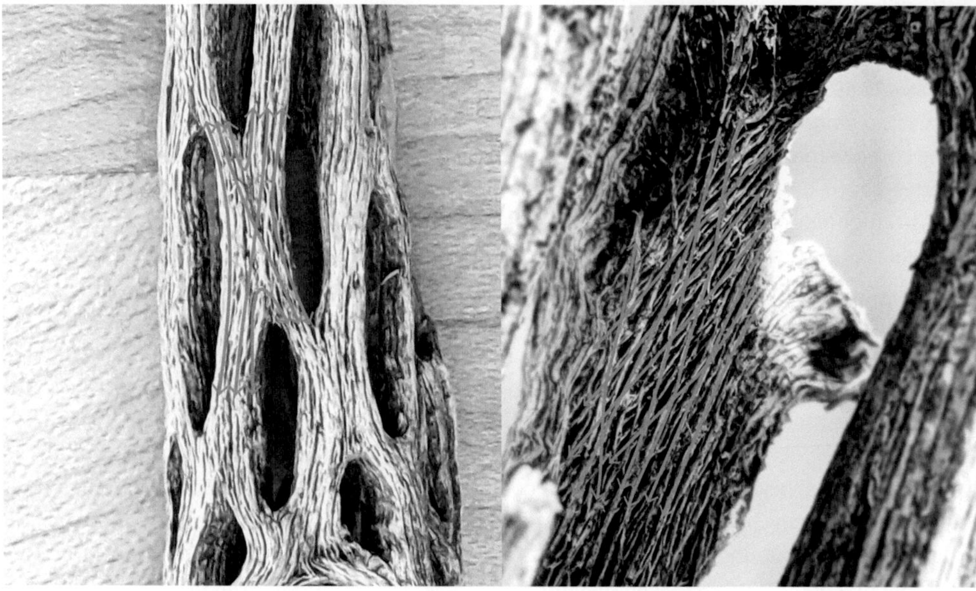

FIGURE 8.6 Biomimetic investigation of a Cholla cactus.

standard lattice construction, which typically connects at the intersections of the lattice. This enabled more continuous material at the intersections of the lattices where structural forces were highest, refined the mid-strut joints with simplified planar connections, and reduced the number of components needed to fabricate the lattice.

The multi-node components were fabricated using an adaptive robotic winding process. The complex and unique forms of the components meant that a standard mandrel mold could not be used, and serialized production processes were not applicable. A reconfigurable frame was developed that could adapt to the variable forms and constraints of the component geometries. This reduced the frame material required for fabrication down to an off-the-shelf kit of parts. The glass and carbon filament were pulled through a resin bath and wound around nodes on the frame. This allows the wet composite material to span between nodes until the resin has cured. Following the logics of the cactus skeleton and the limitations of coreless filament winding, a robotic sequence was developed that would build a composite surface from individually wound fibers. These fibers would densify throughout the process into a robust multi-directional material layout. By controlling the process, the straight fibers would sequentially layer into a smooth hyperbolic surface spanning between the multiple frame nodes. This process was repeated to build the initial form with glass fibers and then reinforce it with carbon fiber per the structural requirements. This created a didactic map of the structure through the material layout that challenges traditional construction concepts of homogeneity in materials. Once cured, the composite material could be removed from the winding frames as lightweight structural units.

Custom 3D-printed components were developed for the base of the tower. Due to the minimized weight of the fiber-wound components, ballast was needed to keep the structure from blowing away, which could also serve as the interface between the ground and the rest of the structure. AM was used to create three heavy base components out of chopped carbon fiber-reinforced Acrylonitrile Butadiene Styrene (CF-ABS). Form and weight were more important than structural efficiency at the base. Each 3D-printed component weighed as much as all the fiber-wound components combined. The form of the base components, which also served as benches for the visitors, could be contoured to fit the curvature of the body. Each bench integrates construction detailing for the assembly of the tower parts and the interface with the fiber-wound composite components (Figure 8.7).

The coreless filament winding and AM processes enabled a large-scale structure with a height of thirty feet to efficiently and effectively be produced from fiber-reinforced plastics. These processes highlighted the potential for plastics to be utilized for architectural applications and showcased the performative opportunities for designing and constructing with integrated material intelligence. Twenty-seven unique fiber-wound components and three unique AM components were produced and assembled. These components minimized material use, maximized structural efficiency, adapted to various architectural constraints, and produced a sustainable production method. Though these processes demonstrated novel opportunities for FRP use in architecture, further investigation was needed to explore hybrid composite manufacturing.

Developing Hybrid Thermoplastic-Composite Components is an ongoing area of exploration that seeks to find synergies between AM and composite production for architectural applications. Both material systems have benefits and limitations for architectural production. Combining AM and composite production into Hybrid Thermoplastic-Composite Components enables design and production opportunities that neither alone can offer. Two prototype systems have been tested

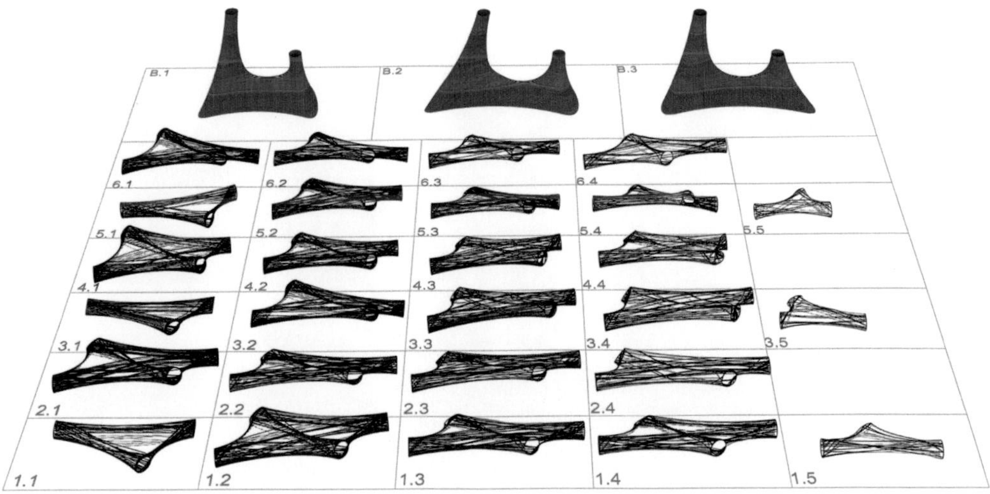

FIGURE 8.7 Non-uniform components produced with adaptive composite manufacturing.

thus far: a hybrid composite sandwich panel that integrates custom 3D-printed ribs that support layers of composite membranes and coreless filament-wound components utilizing continuous fiber thermoplastic extrusion that serves as a substrate for 3D-printed reinforcement fins. In both scenarios the combination of fiber composite production and AM processes augment each other to provide increased design flexibility, geometric freedom, structural performance, and material resourcefulness. These prototypes test lightweight, reinforced, and cantilevering scenarios that are problematic with standard applications of AM in architecture, such as concrete 3D printing.

Several research goals and assumptions were tested to highlight the potential of hybrid composite manufacturing systems. First, material interfaces between printed plastics and composite materials were examined. This included extruding thermoplastics on various composite fabrics, laminating polymer composite fabrics on 3D printed forms with epoxy adhesives and thermal welding, co-extruding continuous filament with thermoplastics, and 3D printing on non-regular wound mesh surfaces. A variety of readily available plastics and composite materials were tested. This included plastics from local waste streams to address the resourcefulness and sustainability of plastic construction. Post-consumer soda bottles made of Polyethylene Terephthalate (PET) ♳, unused 3D printer filament or recycled models made of Polylactic Acid (PLA) ♹, and plastic spools made of Polystyrene (PS) ♸, were ground into pellet sizes suitable for large-format extrusion. The most suitable composite materials for these experiments were polyester sheets, fiberglass weaves, and continuous fiberglass fibers. All of these created a valid interface between extruded plastic and composite materials (Figure 8.8).

AM, composite manufacturing, and robotic fabrication promise increased efficiencies, faster production times, reduced labor, and material savings. Custom tools, fixtures, and processes were developed to accomplish this for the hybrid composite sandwich structures. This includes printing on a convex universal mold, which reduced printing times and material usage for curved shell structures while also allowing a wide variety of geometries to be produced. Custom slicing software was developed to generate continuously curved space-filling ribs for component

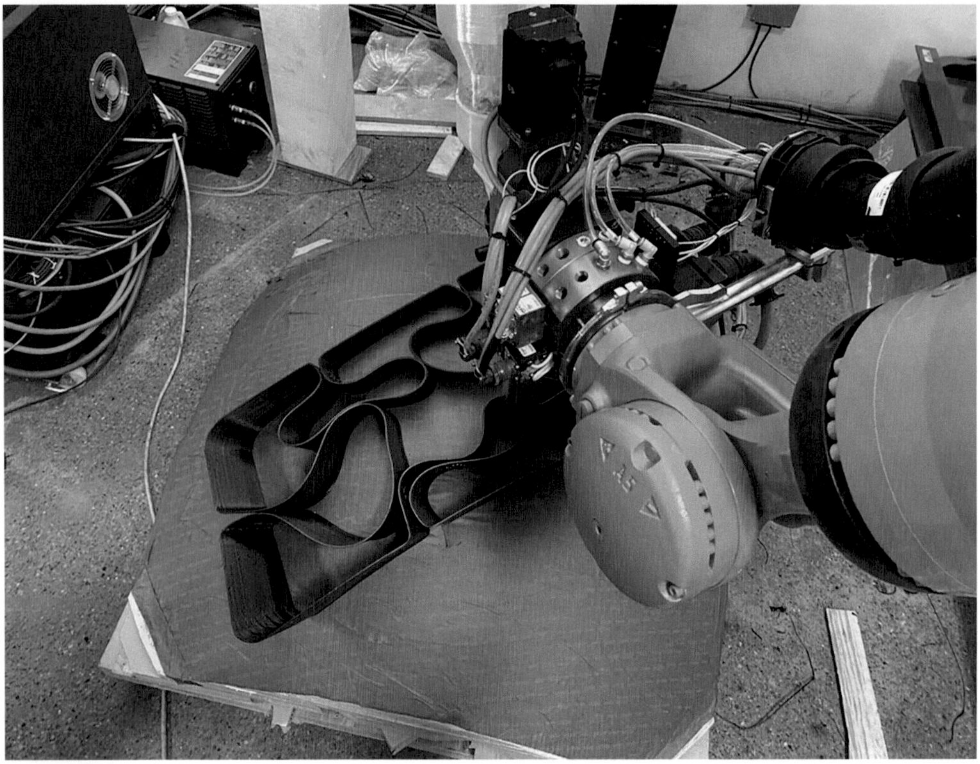

FIGURE 8.8 Hybrid thermoplastic-composite printing.

geometries. This algorithm could be tailored to control material density, span, aperture, curvature, and print speed. The smooth curvature of the continuous print path was considered more efficient for large-format print because the robotic arm could maintain optimal printing speeds around larger radii. Variable height print layers were also integrated to seamlessly blend geometries as component thicknesses changed. The stepped contours produced by standard slicing software were not suitable for laminating composite membranes. Print speed and extrusion rates were adaptively changed to maintain print quality as thicknesses varied (Figure 8.9).

The continuous fiber thermoplastic extrusion process provided tremendous flexibility for hybrid production. Two different fabrication methods could be produced by equipping a large-format extruder with a custom 3D-printed nozzle that could integrate a continuous fiber within an extrusion process. Moving the robot proportionally faster than an extrusion rate meant a tension-based fiber winding process could be achieved. This enabled material to span further than a typical extrusion process alone. By slowing the robotic speed, an extrusion-based process could be utilized. This created the potential to place composite fibers on a surface along non-geodesic paths. This would not be possible with typical coreless filament winding. Both scenarios were used interchangeably to create reinforced hybrid composite components.

FIGURE 8.9 Continuous fiber thermoplastic extrusion processes.

These case studies represent a step towards a hybrid composite manufacturing process that enables the seamless integration of material behavior, structural performance, design freedom, and production efficiency. As Tim Ingold alluded, the combination of these diverse material systems and the adaptive fabrication processes has led to new architectural territory and research potential. These hybrid composite manufacturing processes elevate the value of plastic materials for architectural production and challenge the traditional construction paradigms. They enable the reuse and upcycling of waste materials for architectural applications in a world where the sustainability of building practices is becoming ever more important. They showcase plastics as a material for transformation, not degradation. This provides a vision where "becoming plastic" can create a more adaptive, sustainable, and performative architectural future (Figure 8.10).

Acknowledgement

Text copyedited for clarity and style with Generative AI tools. OpenAI. January 10, 2025.

FIGURE 8.10 Hybrid composite.

Notes

1 "Concrete: The Most Destructive Material on Earth," Jonathan Watts The Guardian, accessed September 30, 2025, www.theguardian.com/cities/2019/feb/25/concrete-the-most-destructive-material-on-earth.
2 Geoff Hammond et al., *Inventory of Carbon & Energy: ICE* 5 (Sustainable Energy Research Team, Department of Mechanical Engineering …, 2008).
3 Tim Ingold, *Making: Anthropology, Archaeology, Art and Architecture* (Routledge, 2013), 18–31.
4 Stephen Phillips, "Plastics: Monsanto Home of the Future," in *Cold War Hothouses: Inventing Postwar Culture, from Cockpit to Playboy*, ed. Beatriz Colomina, Annmarie Brennan and Jeannie Kim (Princeton Architectural Press, 2004) 91–123.
5 Anastasiia Moskaleva et al., "Fiber-Reinforced Polymers in Freeform Structures: A Review," *Buildings* 11, no. 10 (2021), https://doi.org/10.3390/buildings11100481.
6 Fanran Meng et al., "Replacing Plastics with Alternatives Is Worse for Greenhouse Gas Emissions in Most Cases," *Environmental Science & Technology* 58, no. 6 (2024).
7 Marshall Prado et al., "Core-Less Filament Winding: Robotically Fabricated Fiber Composite Building Components," in *Robotic Fabrication in Architecture, Art and Design 2014* (Springer, 2014).
8 Achim Menges and Jan Knippers, *Architecture Research Building: ICD/ITKE 2010-2020* (Birkhäuser, 2020).

References

Hammond, Geoff, Craig Jones, and others. *Inventory of Carbon & Energy: ICE* 5. Sustainable Energy Research Team, Department of Mechanical Engineering …, 2008.

Ingold, Tim. *Making: Anthropology, Archaeology, Art and Architecture*. Routledge, 2013.

Meng, Fanran, Miguel Brandão, and Jonathan M. Cullen. "Replacing Plastics with Alternatives Is Worse for Greenhouse Gas Emissions in Most Cases." *Environmental Science & Technology* 58, no. 6 (2024): 2716–27.

Menges, Achim, and Jan Knippers. *Architecture Research Building: ICD/ITKE 2010-2020*. Birkhäuser, 2020.

Moskaleva, Anastasiia, Alexander Safonov, and Enrique Hernández-Montes. "Fiber-Reinforced Polymers in Freeform Structures: A Review." *Buildings* 11, no. 10 (2021): 481. https://doi.org/10.3390/buildings11100481.

Prado, Marshall, Moritz Dörstelmann, Tobias Schwinn, Achim Menges, and Jan Knippers. "Core-Less Filament Winding: Robotically Fabricated Fiber Composite Building Components." In Wes McGee and Monica Ponce de Leon (Eds.), *Robotic Fabrication in Architecture, Art and Design 2014*, 275–89. Springer, 2014.

Watts, Jonathan. "Concrete: The Most Destructive Material on Earth." Accessed September 30, 2025. www.theguardian.com/cities/2019/feb/25/concrete-the-most-destructive-material-on-earth.

9

NATURE'S BLUEPRINT

Redefining Plastic Designs

Julia Koerner-Al-Rawi

Introduction

In nature, everything is moving. Mountains rise and erode, rivers carve landscapes, and living organisms continuously evolve. This perpetual transformation is the essence of becoming. It is not a final state, but an ongoing journey of adaptation, growth, and redefinition. As designers working with plastic technologies, we embrace this philosophy of becoming, transforming static, mass-produced plastic into dynamic, responsive, and sustainable creations.

Plastics have long been synonymous with permanence and mass uniformity. Their legacy is a testament to industrial success but also environmental consequences. Computational design enables us to combine additive manufacturing and respect for natural morphologies, and we are pioneering a new era in which plastic enables us to mirror nature's intelligence, beauty, and adaptability.

This case study explores how the evolving narrative of plastic reflects a broader metamorphosis in design thinking. Through applied research and fabrication, from a few selected projects, the Lamella Series, the HY and Kelp Clutches, and Cactus Homes, we explore rethinking rigid mass production towards fluid and sustainable creation. We reflect on how this shift empowers designers to take responsibility for environmental stewardship and rethink a future where design is deeply intertwined with the ecosystems from which it draws inspiration.

Rethinking Plastic

For over a century, plastic symbolized human progress. Lightweight, affordable, and versatile, it infiltrates every aspect of modern life. Yet, its very durability led to its downfall. Landfills overflowed, oceans filled with microplastics, and public perception shifted from marvel to menace. The global proliferation of single-use plastic products has underscored the need for radical change in the way materials are designed, used, and understood.

The legacy of plastic, however, is not immutable. We view plastic not as a finality but as raw potential—a material in a state of becoming. A colleague of mine, Manuel Jiménez García, designer and co-founder of NAGAMI, once noted that there is already so much plastic existing

DOI: 10.4324/9781003613589-11

in the world that we must think about how we best recycle and use it, because it is not going to leave our planet.[1] Utilizing robotic technology to give plastic a second life is one way to innovate the reuse of this material. Another example is the 3D printing company AZURE Printed Homes, in California; they utilize the equivalent amount of 100,000 recycled plastic bottles to build 120 sq ft buildings.[2] With JK3D we are leveraging biodegradable polymers and advanced fabrication methods to redefine plastic's narrative from disposable to desirable, from pollutant to performer. This transformation is not merely aesthetic but systemic, suggesting new possibilities for ethical production models that prioritize longevity, recyclability, and biomaterial integration.

Research Methods

Nature presents materials that evolve seamlessly with their environment. Bone, coral, kelp, and mushroom lamellas grow with purpose, responding to stress, light, and gravity. By embracing these natural properties, plastic too can adopt a new identity: one that is alive, adaptive, and inherently sustainable. Through computational design, we infuse plastic with generative intelligence. We work with algorithms which simulate natural growth, creating forms that balance strength, flexibility, and aesthetics. We develop these algorithms either with Rhino Grasshopper or Python Scripting and we also use polygon modelling techniques to 3D model and create geometry which mimic natural growth. While algorithms enable us to recreate realistic mathematical structures, 3D modeling allows for natural deformations and errors and deviations such as they exist in nature. For example, there is no perfect symmetry in nature; however, with computers we can create perfect symmetrical 3D models. This is not necessarily imitation; it is an integration. The structures we develop are informed by biological intelligence, optimizing the relationship between form and performance while echoing natural aesthetics. Such biomimicry designs not only honor organic principles but also reintroduce empathy into material culture.

Biomimicry and Digital Craft: Nature as Blueprint

Biomimicry—the practice of learning from and emulating nature's forms, processes, and systems—offers a powerful model for sustainable and innovative design.[3] Rather than imitating appearances, it seeks to replicate nature's underlying strategies: efficient structures, adaptive behaviors, and closed-loop systems that have evolved over billions of years.

With the rise of computational tools and digital fabrication, biomimicry has found a new ally in digital crafting. Designers can now simulate biological growth patterns, optimize structural performance, and fabricate complex geometries using tools such as parametric modeling, robotic weaving, and 3D printing.[4] These technologies allow us to translate natural intelligence into physical forms that would have been impossible to make by hand.

For example, filament-based fabrication mimics the material logic of bone or plant fibers, distributing material only where structurally necessary—an idea borrowed directly from biological efficiency.[5] Similarly, the branching algorithms used in generative design echo the logic of trees, rivers, and vascular systems.[6]

This convergence of biomimicry and digital crafting reshapes how we design buildings, products, and environments—moving beyond aesthetics toward systems that are performative, responsive, and regenerative.[7] As digital tools make it possible to work at the scale and precision of nature, designers are empowered to craft solutions that are both technologically advanced and ecologically attuned.

Fabrication Process

The underside of a portobello mushroom reveals nature's mastery of structure, the Hymenium. Lamellas, delicate yet strong, distribute weight efficiently while enabling spore dispersal. Initially we designed a 3D printed wearable in 2015 which was inspired by the structure of fungi, the Hymenium Jacket, part of the Sporophyte collection and the first ready-to-wear collection in the United States. We 3D scanned the mushroom with open-source 3D scanning software on our smart phone and thereafter developed an algorithm in Rhino Grasshopper which mimicked the lamella undulation of the Hymenium. We transformed the pattern into a 3-D surface model of the wearable and 3D modeled the geometry with thickness, undulations, and irregularities to make it look more naturally grown. We then 3D printed the design with Stratasys' Polyjet technology. PolyJet is a high-resolution additive manufacturing process that works by jetting photopolymer droplets onto a build platform, which are immediately cured using ultraviolet (UV) light. This technology enables the printing of smooth, highly detailed parts with multiple materials and colors in a single build, making it ideal for prototyping complex designs, medical models, and product mockups. Its ability to simulate rubber, transparent, and rigid materials in a single print sets it apart from other 3D printing methods.[8]

Inspired by this, our Lamella Series, developed in 2021, reinterprets these principles through additive manufacturing in a series of products which include pedestals and vases of many sizes. Each object is printed as a continuous thin shell, optimizing material use and eliminating the need for support structures. The process only uses the amount of material which is needed to produce one object.

The rippling surfaces, echoing natural topologies, enhance structural performance while reducing mass. Variations in algorithmic parameters ensure every piece is unique, echoing the natural world's infinite diversity. This exploration allows for a poetic translation of ecological forms into objects that challenge conventional design norms, standing as functional sculptures with ecological intelligence.

The Lamella Series is 3D printed with biodegradable plastic polylactic acid (PLA) in various colors, locally sourced and produced on demand. PLA is a biodegradable thermoplastic derived from renewable resources such as corn starch, sugarcane, or cassava. It is one of the most widely used bioplastics, particularly popular in 3D printing due to its low melting point, ease of processing, and environmentally friendly profile. Unlike petroleum-based plastics, PLA decomposes under industrial composting conditions, making it a more sustainable alternative in applications such as packaging, disposable items, and additive manufacturing.[9]

We utilize desktop Filament Deposition Modeling (FDM) 3D printers for the Lamella Series; in Vienna, Austria, we have a large FDM printer which we utilize to print larger-scale designs up to 100 cm in height and 50 cm in diameter. The various sizes can be stacked to architectural columns and nested as a family of objects (see Figures 9.1 and 9.2).

Fused Deposition Modeling (FDM) is an additive manufacturing technique in which a thermoplastic filament—commonly PLA or Acrylonitrile Butadiene Styrene (ABS)—is heated to a semi-liquid state and extruded through a nozzle to build objects layer by layer. Known for its simplicity, cost-effectiveness, and accessibility, FDM is widely used in prototyping, product development, and education.[10] Despite some limitations in surface finish and resolution compared to other 3D printing methods, FDM remains one of the most popular technologies in desktop and industrial applications.

FIGURE 9.1 Lamella Series closeup.

FIGURE 9.2 Lamella Series.

Applications in Design and Architecture

The Crystal Lamellas: Fragility and Strength in Glass

Expanding our exploration of lamella systems, the Crystal Lamellas Art Piece pushes material boundaries. Exhibited during the Italian Glass Weeks 2022 at the Gritti Palace, a Luxury Collection Hotel in Venice, this collection translates natural ripples into crystalline formations. Glass, traditionally seen as fragile, assumes new strength through geometric intelligence (see Figure 9.3).

3D Printed in collaboration with Austrian crystal glass company Swarovski, these structures merge algorithmic precision with artisanal craftsmanship, utilizing their innovative novel 3D printing glass technology and individually coated with their signature colorization technology. Each of the 40 vessels is 3D-printed spirally as a single shell with only the necessary material of glass. The play of light across their surfaces recalls the iridescence of deep-sea creatures and mineral crystals, embodying the poetic synergy of natural inspiration and digital fabrication. The delicate relationship between geometry and translucency creates dynamic interactions with space and illumination, extending the expressive potential of glass as both material and medium. The technology Swarovski has developed over the past years is innovative as they adapted Fused Filament Fabrication (FFF) principles to process glass-based materials. In this approach, a filament composed of glass particles is extruded layer by layer through a heated nozzle, forming intricate crystal-like structures. This adaptation of FFF enables Swarovski to digitally fabricate complex, customized crystal components with precision and repeatability, bridging traditional

FIGURE 9.3 Crystal Lamellas.

artisanal craftsmanship with cutting-edge additive manufacturing. The process offers new creative possibilities for producing detailed, lightweight, and sustainably made glass elements, while maintaining the hallmark brilliance and purity of Swarovski's crystals.

Alongside the Vessels, we designed plinth-like trays for the art pieces which are 3D printed with natural quartz sand in a binder jetting technology, manufactured by the company Sandhelden in Germany. The combination of glass and sand 3D printed objects tells a story on their own about material and transformation.

The HY Clutch and Kelp Mini Clutch: Digitally Grown Elegance

In the ever-evolving landscape of fashion and product design, our vision of the future is where technology, sustainability, and nature converge. With the HY Clutch we pioneered the cutting-edge 3D-printed fashion movement, creating limited, on-demand collections that are as beautiful as they are sustainable (see Figure 9.4). Released in early 2021, the handbag is a marvel of digital design and biomimetic inspiration. Similarly to the Lamella Series, the design is drawn from the microscopic world of fungi, specifically the lamellar structures of the hymenium, the gill-like tissue beneath mushroom caps. Its outer surface is composed of hundreds of delicate lamellae, layered to create structural integrity and dynamic translucency. The design is precise and intricate and complex. It can only be realized through 3D printing. The handbag, composed of 3,626 layers, is fully 3D printed—from its hinge and clasp to its sculpted interior. It demonstrates the unmatched capacity of additive manufacturing to integrate complex geometry and function into a seamless whole. The interplay of light and form produces

FIGURE 9.4 Lamella Series and HY clutches.

a unique transparency that allows a glimpse into its interior, further reinforcing the clutch's visual lightness. Made from bio-based polymers derived from renewable resources like soy and corn, the clutch reduces reliance on fossil fuels and toxic materials.

We utilize Stereolithography (SLA), a resin-based 3D printing technology that uses an ultraviolet (UV) laser to selectively cure liquid photopolymer resin, solidifying it layer by layer to build a 3D object. SLA is known for its exceptional resolution and surface finish, making it ideal for applications requiring fine detail, such as jewelry, and highly accurate prototypes. Unlike FFF, which extrudes solid material, SLA relies on light-induced polymerization in a vat of photosensitive resin. SLA's advantages include high precision, smooth surface quality, and the ability to produce complex geometries with minimal post-processing. However, it typically requires support structures, post-curing, and careful handling of liquid resins, which may be brittle or sensitive to UV exposure over time.[11]

The handbags are manufactured in-house in our ateliers. The decision to produce locally and on-demand reflects our commitment to minimizing waste and reducing carbon emissions, setting an example of how accessories can be created responsibly. This design was honored with the prestigious German Red Dot Product Design Award in 2021, a recognition of its innovation, aesthetic value, and sustainability.

Where the HY Clutch explores the architecture of fungi, the Kelp Mini draws its inspiration from the undulating forms of marine life. Specifically, we were inspired by kelp specimens collected and 3D-scanned in Malibu, California. The resulting design embodies the rhythm and intricacy of underwater ecosystems. Its flowing geometries and sculptural voids allow for weight

FIGURE 9.5 Kelp Mini.

reduction and light diffusion, creating an ethereal visual effect that mimics sunlight filtering through ocean plants. With the kelp motif appearing in earlier JK3D works like the Kelp Mask, Kelp Necklace, and Kelp Jacket, the clutch represents a mature evolution of our organic design language. The necklace, part of the permanent collection at the Museum of Applied Arts (MAK) in Vienna, speaks to the significance and cultural value of our designs.

Lightweight yet strong, the Kelp Mini maximizes the structural benefits of additive manufacturing (see Figure 9.5). Its internal cavity is practical for everyday use, while the outer latticework exudes elegance. As with the HY Clutch, the Kelp Mini is made from biodegradable plant-based polymers and fabricated locally in Los Angeles and Vienna using renewable energy. We also spend a significant amount of time optimizing the printing process and reducing support structures necessary for the resin printing process. This optimization has led to reduced material waste during production and is only possible with additive manufacturing. It exemplifies how material consciousness can be woven seamlessly into design. We are revolutionizing the lifecycle of fashion accessories. Each piece is printed only when ordered, eliminating overproduction and the need for warehousing. The clutches are also designed to be personalized, strengthening emotional attachment, and encouraging long-term ownership (see Figure 9.6).

The packaging is reimagined; we design a custom protective case made from recyclable plastics, echoing the clutch's own geometry. It is a full-circle approach, where even transport and storage reflect our studio's values of form, function, and sustainability. We challenge traditional notions of luxury, production, and design, rethinking how we create fashion, and we offer a vision that is ethical, digitally crafted, and ecologically informed.

FIGURE 9.6 Hy Mini Schloss collection.

The Cactus Homes: Architecture Rooted in Adaptation

In arid environments, cacti thrive through structural ingenuity and water conservation. The award-winning Cactus Homes architecture project, developed between 2023 and 2024, extrapolates these lessons to reimagine affordable, sustainable housing (see Figure 9.7). Designed for Texas-based ICON's large-scale concrete 3D printing technology, we designed homes with ribbed, porous surfaces that enhance thermal regulation and structural integrity. ICON's Vulcan system is a large-scale, gantry-based 3D printer designed to construct buildings by extruding Lava Crete, a proprietary cementitious material, layer by layer. Controlled by Build OS software, the system translates architectural designs into print-ready toolpaths, enabling rapid, automated construction directly on site. ICON's integrated Magma system prepares and delivers Lava Crete, adapting in real time to environmental conditions. This method significantly reduces labor and material waste while enabling complex geometry and durable structures. ICON has used this technology to build entire communities, military barracks, and is collaborating with NASA on lunar habitat prototypes.[12]

As part of the Initiative 99, we designed houses which fit a 2-unit duplex configuration. The layout enhances the efficiency, functionality, and customizability of the homes, while still following the principles of Community First! Village, a sustainable housing model focused on fostering social connections and environmental responsibility. The duplex units incorporate features such as operable skylights, integrated storage, and flexible room dividers, ensuring that each home remains adaptable to the unique needs of its residents, all while maintaining a focus on sustainability and affordability.

In the development of the Cactus Homes design, we also collaborated with Altair software company. We sent 3D files, and they would perform structural analyses for us. They mainly conducted a series of structural simulations in Altair's Simsolid software to benchmark the

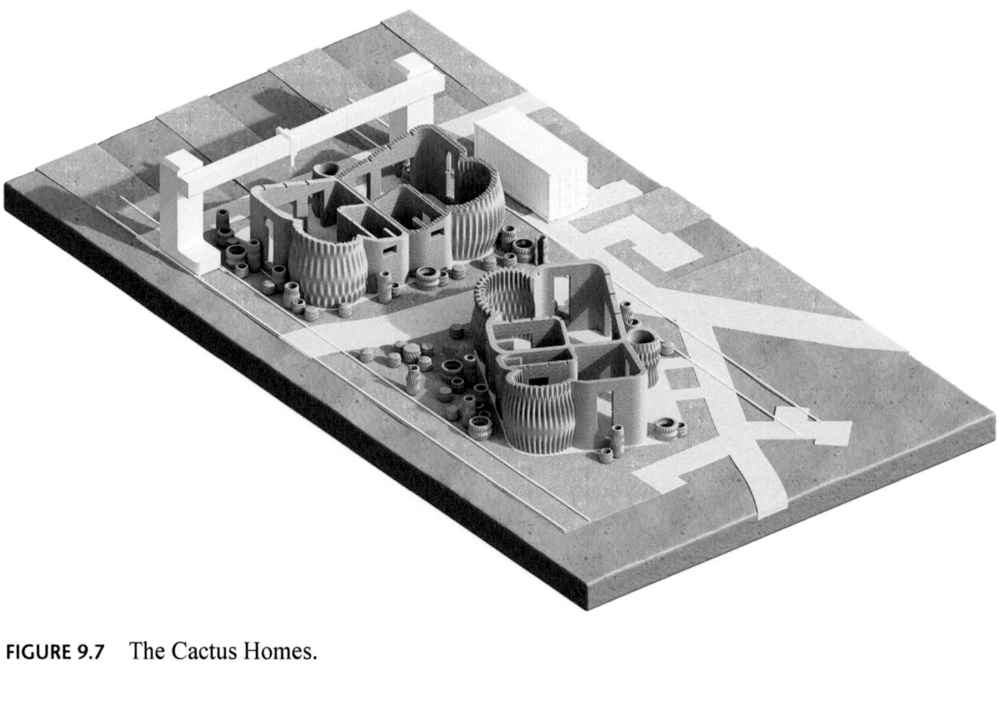

FIGURE 9.7 The Cactus Homes.

house's geometry against other simple curvilinear forms—which can already be efficient—yet we wanted to demonstrate how organic forms can potentially do more with less. In other words, they were testing and validating different design variations for the living spaces, walls, and water tanks. The goal was to lower stresses and deflections and minimize material volume and weight. The results ensured we were getting the best, most efficient designs possible and adapting our workflows to generate more efficient forms.[13]

Algorithmic design allows for parametric variation, ensuring each home responds to its specific environmental context. Like their botanical counterparts, these structures are both resilient and resource-efficient, marking a pivotal step in scaling biomimicry from product to architecture. In this endeavor, domestic space becomes a site for experimentation in sustainability, resilience, and adaptation—qualities increasingly urgent in the face of climate change.

Conclusion

Algorithmic Design: From Code to Craft

At the heart of our practice lies algorithmic design. Generative scripts function as digital DNA, encoding patterns of growth, stress distribution, and material efficiency. Rather than imposing form, we work with algorithms developed mainly in Rhino Grasshopper as well as polygon modeling software to evolve shapes organically, much like nature sculpts natural artifacts like bone and coral. This methodology transcends aesthetics; it is an ethical imperative. Designs are not static blueprints but living systems, capable of endless variation without additional production costs or environmental burden. Such flexibility enables decentralized, localized production models and empowers communities to customize designs in response to cultural, climatic, and contextual needs. 3D printing empowers us to fabricate directly from these generative models, bypassing the constraints of traditional molds and mass production. We produce on demand, reducing inventory and material waste. Biodegradable filaments and recycled materials further diminish environmental impact. Through this seamless digital-to-physical workflow, we achieve unprecedented freedom. Our design processes resist the norms of replication and sameness. Instead, they embrace the richness of variation as a design asset, forging a new path for the aesthetics of sustainability.

Material Futures: Expanding the Palette beyond Plastic

While plastic remains central to our exploration, our vision extends to alternative materials. Crystal Lamellas demonstrates the poetic potential of glass, while Cactus Homes embodies the promise of concrete on an architectural scale. Each material carries distinct challenges and opportunities for sustainable design. These material evolutions reflect our commitment to scalability and sustainability. Whether crafting intimate objects or inhabitable spaces, we apply the same principles: minimal waste, maximum aesthetic value, and structural intelligence drawn from the natural world. By continuously testing the limits of material and digital design processes, we open new avenues for circular design models and interdisciplinary innovation.

Hybrid Processes and Sustainable Craftsmanship

Our atelier operates as a hybrid laboratory, merging digital precision with artisanal sensibility. The solar-powered facility in Los Angeles and the renewable-energy-driven facility in

Vienna enable clean-energy fabrication, while localized production reduces shipping emissions. The integration of biodegradable polymers, plant-based filaments, and recycled aggregates exemplifies our material ethic. We envision a future where manufacturing is not extractive but regenerative—where every object is born from ecological consideration. This regenerative design ethos challenges the disposable culture and fosters emotional durability; encouraging users to cherish and preserve the objects they interact with, while offering alternative solutions for reusing materials for instant re-creation. Wouldn't it be great if you can turn your recycled glass instantly into a new object or melt your clothing to 3D print new ones "at home"?

Co-Creation and Collaboration: A Multidisciplinary Approach Designers, Technologists, and Educators

True innovation thrives at the intersection of disciplines. Our collaborations span computational designers, material scientists, engineers, and master craftspeople. Each project becomes a dialogue between digital possibility and material reality. This dynamic interplay fosters innovation, enabling new typologies, processes, and aesthetics. This ethos is exemplified in Crystal Lamellas, where algorithmic design converges with crystal glass expertise, and in Cactus Homes, where digital fabrication aligns with ecological architecture. Together, these projects form a design ecosystem—responsive, inclusive, and forward-looking. We are committed to education and dissemination. Workshops, exhibitions, and research publications share our insights with a global audience, nurturing a community of like-minded creators dedicated to sustainable design evolution. By making our methodologies transparent, we invite others to contribute, reinterpret, and apply them in diverse contexts.

In the Art of Becoming, Plastic Finds its Evolution

Plastic becomes not an environmental adversary but an ecological ally. Through algorithmic intelligence, sustainable materials, and nature-inspired design, we redefine what plastic can be. Projects like the Lamella Series, Hy and Kelp Clutch, and Cactus Homes embody this transformation. They are not endpoints but milestones in an ongoing journey of becoming a testament to the power of design to align material innovation with ecological wisdom.

As we project into the future of digital craftsmanship, our commitment is to craft objects and environments that honor the intelligence of natural systems, harness the precision of digital fabrication, and contribute to a regenerative future. We see design not just as a profession, but as a responsibility, a way to imagine and build futures that are symbiotic with the living world. In the art of becoming, plastic finds its evolution.

Acknowledgements

I, as the author of this book chapter, acknowledge the use of the generative AI tool ChatGPT (GPT-5) to edit and refine the English language of the text after I had written it. The tool was used solely for language polishing and minor stylistic adjustments, as my native language is German.

Notes

1 Invited Guest Lecture as part of Research Studio at UCLA AUD on 04.13.2023 https://nagami.design/en.
2 "Azure Printed Homes," Azure Printed Homes, accessed June 19, 2025, www.azureprintedhomes.com/.
3 Janine M. Benyus, *Biomimicry: Innovation Inspired by Nature* (Harper Perennial, 2002).
4 Achim Menges and Jan Knippers, "Fibrous Tectonics: Digital Design and Fabrication of Filament-Based Construction," *Architectural Design* 85, no. 5 (2015), https://doi.org/10.1002/ad.1952; Paola Antonelli et al., *The Neri Oxman Material Ecology Catalogue* (The Museum of Modern Art, 2020).
5 Julian F. V. Vincent et al., "Biomimetics: Its Practice and Theory," *Journal of the Royal Society, Interface* 3, no. 9 (2006), https://doi.org/10.1098/rsif.2006.0127.
6 Michael Hensel et al., *Emergent Technologies and Design: Towards a Biological Paradigm for Architecture* (Routledge, 2010).
7 Michael Pawlyn, *Biomimicry in Architecture* (RIBA Publishing, 2011).
8 Gideon N. Levy et al., "Rapid Manufacturing and Rapid Tooling with Layer Manufacturing (LM) Technologies, State of the Art and Future Perspectives," *CIRP Annals* 52, no. 2 (2003), https://doi.org/10.1016/S0007-8506(07)60206-6.
9 L.-T. Lim et al., "Processing Technologies for Poly(Lactic Acid)," *Progress in Polymer Science* 33, no. 8 (2008), https://doi.org/10.1016/j.progpolymsci.2008.05.004.
10 Brian N. Turner et al., "A Review of Melt Extrusion Additive Manufacturing Processes: I. Process Design and Modeling," *Rapid Prototyping Journal* 20, no. 3 (2014), https://doi.org/10.1108/RPJ-01-2013-0012.
11 Tuan D. Ngo et al., "Additive Manufacturing (3D Printing): A Review of Materials, Methods, Applications and Challenges," *Composites Part B: Engineering* 143 (2018), https://doi.org/10.1016/j.compositesb.2018.02.012.
12 "Technology," ICON Technology, Inc., accessed June 19, 2025, www.iconbuild.com/technology.
13 "Expert Q&A: How Julia Koerner Used Altair SimSolid to Design Sustainable, Affordable Housing," Altair, accessed June 19, 2025, https://altair.com/blog/articles/how-julia-koerner-used-altair-simsolid.

References

Altair. "Expert Q&A: How Julia Koerner Used Altair SimSolid to Design Sustainable, Affordable Housing." Accessed June 19, 2025. https://altair.com/blog/articles/how-julia-koerner-used-altair-simsolid.

Antonelli, Paola, Anna Burckhardt, and Hadas A. Steiner. *The Neri Oxman Material Ecology Catalogue*. The Museum of Modern Art, 2020.

Azure Printed Homes. "Azure Printed Homes." Accessed June 19, 2025. www.azureprintedhomes.com/.

Benyus, Janine M. *Biomimicry: Innovation Inspired by Nature*. Harper Perennial, 2002.

Hensel, Michael, Achim Menges, and Michael Weinstock. *Emergent Technologies and Design: Towards a Biological Paradigm for Architecture*. Routledge, 2010.

ICON Technology, Inc. "Technology." Accessed June 19, 2025. www.iconbuild.com/technology.

Levy, Gideon N., Ralf Schindel, and J. P. Kruth. "Rapid Manufacturing and Rapid Tooling with Layer Manufacturing (LM) Technologies, State of the Art and Future Perspectives." *CIRP Annals* 52, no. 2 (2003): 589–609. https://doi.org/10.1016/S0007-8506(07)60206-6.

Lim, L.-T., R. Auras, and M. Rubino. "Processing Technologies for Poly(Lactic Acid)." *Progress in Polymer Science* 33, no. 8 (2008): 820–52. https://doi.org/10.1016/j.progpolymsci.2008.05.004.

Menges, Achim, and Jan Knippers. "Fibrous Tectonics: Digital Design and Fabrication of Filament-Based Construction." *Architectural Design* 85, no. 5 (2015): 40–47. https://doi.org/10.1002/ad.1952.

Ngo, Tuan D., Alireza Kashani, Gabriele Imbalzano, Kate T. Q. Nguyen, and David Hui. "Additive Manufacturing (3D Printing): A Review of Materials, Methods, Applications and Challenges." *Composites Part B: Engineering* 143 (2018): 172–96. https://doi.org/10.1016/j.compositesb.2018.02.012.

Pawlyn, Michael. *Biomimicry in Architecture*. RIBA Publishing, 2011.

Turner, Brian N., Robert Strong, and Scott A. Gold. "A Review of Melt Extrusion Additive Manufacturing Processes: I. Process Design and Modeling." *Rapid Prototyping Journal* 20, no. 3 (2014): 192–204. https://doi.org/10.1108/RPJ-01-2013-0012.

Vincent, Julian F. V., Olga A. Bogatyreva, Nikolaj R. Bogatyrev, Adrian Bowyer, and Anja-Karina Pahl. "Biomimetics: Its Practice and Theory." *Journal of the Royal Society, Interface* 3, no. 9 (2006): 471–82. https://doi.org/10.1098/rsif.2006.0127.

10

PLASTICS PLUS

Virginia San Fratello

Emerging Objects frequently explores the use of waste materials as sources for 3D printing. Among the most widely consumed beverages worldwide are coffee, tea, and wine, each of which has played a significant role in the history of colonization. Beginning in the 15th century, European settlers introduced coffee trees, tea shrubs, and grapevines to regions such as India, Australia, and the Americas. This large-scale transplantation of plants represents one of the earliest examples of mass-assisted migration in history. Due to the global expansion of these crops and the extensive industries built around them, they now generate enormous amounts of agricultural by-products each year. These agricultural by-products can be compounded with bio-plastics, such as PLA, to create new composite materials that are not only recycled but have embedded cultural values, familiar scents, new haptics and textures, and familiar colorways that are informed by the waste materials themselves bringing added value to the plastic itself, which can be understood as a plus.

Plastics Plus Agricultural and Food Waste

In coffee production, waste is generated at multiple stages, beginning in the fields when coffee cherries are harvested. The coffee bean itself is located at the center of the cherry, which undergoes processing—either drying or washing—to separate the bean from the outer pulp and skin. While the beans continue through the supply chain for roasting, the discarded cherry pulp often ends up in landfills or pollutes waterways. It is estimated that the global coffee industry generates 20.8 billion kilos of coffee cherry waste per year.[1] One innovative solution to repurposing this by-product is coffee flour, created by drying and grinding the discarded coffee cherry into a fine powder that can be used in baking.

A prime example of upcycled coffee waste is the Coffee Coffee Cups, which are 3D-printed using a mixture of coffee flour, maltodextrin, and powdered sugar in a ratio of 4:1:1 by weight. The ingredients are combined in the build bed of a ZCorp 310 Plus binder jet printer, where a fine layer of flour mix (0.127 mm thick) is saturated with a rice wine-based binder. This process results in an edible, functional coffee cup with a pastry-like consistency.

DOI: 10.4324/9781003613589-12

FIGURE 10.1 3D-printed Coffee Coffee Pot.

Another major source of coffee waste comes from the industrial production of instant coffee. The process involves brewing large quantities of coffee and then dehydrating it to create the final product, leaving behind massive amounts of spent coffee grounds. Similarly, coffee shops, restaurants, and households worldwide generate significant quantities of used grounds daily. The Coffee Coffee Pot is an example of how this waste can be repurposed—3D-printed entirely from recycled coffee grounds (Figure 10.1).

Both the Coffee Coffee Cups and the Coffee Coffee Pot were digitally designed using Modo software before being 3D-printed. The resulting pieces possess a deep umber color that darkens over time, developing a patina. While the cups emit a subtle herbal aroma due to their unroasted coffee cherry composition, the pot continuously releases the rich, bitter scent of roasted coffee. Research suggests that the scent of coffee can trigger alertness, and recent studies even indicate that its 28 potent odor compounds may help reduce stress in animals.[2] If materials like coffee flour or recycled grounds are used to construct objects that gradually release these aromatic compounds, it raises the possibility that such objects could have a positive psychological effect on humans. The scent of these coffee-based designs lingers for over a year, subtly influencing the atmosphere of the space they inhabit.

Similar to coffee, the tea industry generates a considerable amount of waste throughout cultivation, processing, and consumption. When tea is harvested, mature leaves, stems, and stalks are discarded in the fields. Additional waste accumulates during refining and packaging,

FIGURE 10.2 3D-printed Utah Tea Set.

with fine tea dust—known as "tea fluff"—collecting on factory floors. Furthermore, like instant coffee production, instant tea manufacturing produces large volumes of brewed and extracted tea solids that are typically discarded. Tea shops and households also contribute to the waste stream, disposing of billions of pounds of used tea leaves each day.

For 3D printing, a specialized mixture is developed using one part tea solids, nine parts refined sugar, and one part maltodextrin. This blend is finely ground, sifted, and placed into the build bed of a binder jet printer. Due to the properties of the mixture, it requires a low saturation rate and a high layer thickness (0.203 mm) to prevent excessive caking during the printing process (Figure 10.2).

Wine production also leaves behind a significant amount of organic waste in the form of pomace, the solid remains of grapes after they are pressed for juice. Pomace consists of grape skins, seeds, stems, and pulp. Historically, this by-product has been used in the production of brandies like Grappa, a tradition dating back to medieval times in Italy.[3] However, even after distillation, substantial amounts of pomace remain. While some of it is repurposed as animal feed or fertilizer, a large portion goes unused. In California alone, wineries generate over 100,000 tons of pomace annually—approximately one ton of waste for every five tons of crushed grapes.[4]

One innovative approach to reusing this material is demonstrated in the Chardonnay Wine Goblets, which are 3D-printed using upcycled Chardonnay grape skins and seeds (Figure 10.3). The pomace is sourced from vineyards in Sonoma County, kiln-dried, and ground to a fine particle size (200 microns). It is then blended with wheat paste in a 6:4 ratio by weight before being printed. These goblets offer a sensory experience, as both the wine they hold and the

FIGURE 10.3 3D Chardonnay Wine Goblets.

material they are made from can be evaluated for their color, texture, aroma, and body. The goblets themselves retain the characteristic pineapple and herbaceous notes of an unoaked Chardonnay.

Not only can food be recycled for 3D printing but also plastic food packaging waste can be recycled for 3D printing. Food plastic waste is a major contributor to environmental pollution, but innovative recycling methods are transforming this waste into valuable resources. By collecting and processing discarded plastic packaging—such as bottles, containers, and wrappers—these materials can be shredded, melted, and extruded into high-quality 3D-printing filament. This approach not only diverts plastic from landfills and oceans but also provides a sustainable alternative to traditional petroleum-based filaments. Using recycled plastic for 3D printing opens up new possibilities for creating durable, functional objects while promoting a circular economy.

The Turmeric Tower is crafted from recycled food packaging waste and naturally dyed in a rich turmeric bath, giving it a vibrant golden hue (Figure 10.4). The Coffee Coffee Bonbonniere is 3D-printed using a sustainable brown filament composed of repurposed coffee grounds and recycled food packaging waste. The Pomegranate Pot is made from an innovative filament derived from recycled pomegranate seeds, showcasing both sustainability and creative material innovation. These objects are all made of polylactic acid, more commonly known as PLA or bioplastic, plus recycled food.

FIGURE 10.4 3D-printed Turmeric Tower.

Plastics Plus Light

3D-printed plastics can also be combined with light to create novel architectural and interior applications which can be seen in the Furry Forest and Star Lounge (Figure 10.5). The Furry Forest is composed of a collection of 3D-printed furry light fixtures designed to evoke joy and delight. During the COVID-19 pandemic, fear of touch became a part of daily life, and many

FIGURE 10.5 Furry Forest.

experienced the darkness of loss and isolation. These luminous, playful totems stand as a bold response to that fear—radiating warmth, inviting interaction, and quite literally begging to be petted.

Each piece is crafted from a sustainable blend of recycled sawdust, PLA, repurposed food packaging waste, and natural agricultural by-products like pomegranate seeds. Every totem is enveloped in 3D-printed "fur," tapping into the innate comfort of soft textures. Hair and fur are biologically linked to sensations of touch and connection, offering pleasure to both the giver and the receiver. The shaggy surface of these light fixtures enhances a sense of domestic comfort and security, making them as emotionally engaging as they are visually striking.

The bioplastic elements burst with bold, high-energy colors, while the wood-based filament introduces an organic warmth, reinforcing a connection to nature. Together, these elements create a dynamic balance between vibrancy and harmony—transforming each individual piece into a glowing, tactile beacon of happiness.

The Star Lounge is another glowing, tactile beacon; Emerging Objects worked with Bold Machines to unveil one of the largest PLA-based 3D-printed structures ever built in 2015. This groundbreaking project showcases the architectural possibilities of 3D printing, particularly through the use of small-scale Desktop 3D printers.

Star Lounge is a freestanding dome that stands 8.5 feet tall with a footprint of approximately 11 by 11 feet (Figure 10.6). Constructed from 2,073 hexagonal blocks, each printed in different

FIGURE 10.6 Star Lounge.

translucent bioplastic colors to indicate block type, the design simplifies assembly while forming a stunning pattern of stars and hexagons across 21 prefabricated panels that are riveted together. The Star Lounge serves as proof that prefabricating 3D-printed panels for architectural applications is not only feasible but also paves the way for 3D-printed walls, ceilings, partitions, and cladding in future construction projects.

The structure was fabricated using a 'BotFarm' of over 100 3D printers, the team optimized print efficiency—each printer producing two blocks in just over an hour without the need for support materials. To streamline construction, only 28 unique block types were used in the doubly curved dome, with each block numbered on the interior for precise placement. Pre-printed rivet holes further simplified assembly.

The final structure creates an intricate, decorative pattern reminiscent of traditional American quilts and Islamic geometric motifs, enveloping visitors in an immersive, luminous almost otherworldly atmosphere. The Star Lounge demonstrates the ability to create lightweight structures with affordable desktop printers.

Plastics Plus Wildfire Ash

The *WUIware* (woo-eee wehr) vessels are 3D-printed from a custom bio-filament made of California wildfire ash collected from the WUI. The WUI, wildland-urban interface, is the area of transition between unoccupied land and human development. It is the zone where buildings and other human development intermingle with undeveloped wildland and many wildfires around the world occur in this zone. Forests, shrubland, and grassland cover more than half of the land area in the United States.[5] These ecosystems are important resources, both environmentally and economically. Although wildfires occur naturally and are essential to a long-term role in their health, changing wildfire patterns, as an outcome of global warming and climate change, threaten to upset these norms. Multiple studies have found that climate change has already led to an increase in wildfire season length, frequency, and burned area.[6] Since 1983, the National Interagency Fire Center has documented approximately 70,000 wildfires per year in the United States. The area burned by wildfires has increased since the 1980s, when temperatures nationally and globally began to increase continuously and noticeably.[7]

As population growth continues, and people move out of cities and into the WUI, or wildland-urban interface, the increase in wildfires means that we also have an increase in severe wildfire risk in these semi-urbanized zones. This translates to greater destruction of homes and building structures, which raises the question, how do we rebuild for a safer future, and can we build using wildfire waste? Could we rebuild using the ashes from the wildfires themselves? Can wildfire waste be used in conjunction with emerging technologies such as 3D printing to create bio-digital materials?

The ashes for the *WUIware* were collected during 2020 from three different California wildfires. The most material was collected from the CZU and the SCU lightning complex fires which burned 480,000 acres and over 1700 buildings. The ash was passed through a 200-mesh sieve and only the finest particles were used to create the bio-filament. The ash was mixed with PLA at ratios of 10%, 20%, and 30%. The 10% and 20% were able to be extruded into a round filament using a twin screw setup. The 30% was too brittle and broke. The 10% and 20% ash filament were tested in a Prussa MK3 desktop printer using 0.4, 0.6, and 0.8 diameter nozzles. The 0.4 and 0.6 mm nozzles frequently clogged due to the impurities in the ash filament. The 20% filament regularly broke while being loaded into the hot end of the printer. The 10% ash filament performed the best most of the time with the 0.8 mm diameter nozzle. The 10% ash filament with the 0.8 mm diameter nozzle was used on the Prussa MK3 to 3D-print the final *WUIware* objects. The bases are printed using a continuous flow design. The filament tended to clog the nozzle when 3D printing designs that required retraction, slowing the 3D printing process.

The outcome of the research resulted in three 3D-printed vessels that are called *WUIware* (Figure 10.7). They are essentially tableware made of the ash from the WUI. The vessels are objects intended to help us remember what we have lost as they are quite literally made of the ash of the trees, homes, and personal artifacts that were burned in the fires and serve to demonstrate that the by-product of the wildfire, in this case, ash, a material of destruction, can be used with 3D printing, to become a material of construction.

Conclusion

The work of Emerging Objects demonstrates the transformative potential of 3D printing when combined with biopolymers, recycled materials, and natural by-products. By repurposing agricultural waste from coffee, tea, and wine production, recycling plastic food packaging, and even integrating wildfire ash into printable materials, these projects push the boundaries of sustainability in digital fabrication.

FIGURE 10.7 WUIware.

From the Coffee Coffee Cups and Chardonnay Wine Goblets to large-scale architectural installations like Star Lounge, these innovations showcase how waste can be reimagined as a valuable resource for design and construction. The Furry Forest and WUIware vessels go beyond aesthetics, engaging with social and environmental narratives—whether by responding to the emotional impact of the COVID-19 pandemic or confronting the devastating realities of climate change and wildfires.

At its core, this research underscores that 3D printing is not merely a tool for fabrication but a catalyst for material innovation and sustainability. By integrating waste streams into architectural, functional, and decorative objects, these projects challenge conventional ideas of what materials can be and how they can positively shape the future of design, architecture, and environmental responsibility. As technology advances, the potential for bio-digital materials to reduce waste, enhance circular economies, and create resilient structures continues to expand—paving the way for a more sustainable and imaginative future in 3D printing.

Notes

1 "Upcycling Coffee Cherries for Food: "Every Major Chocolate Company Is Looking at This," Flora Southey, accessed September 26, 2025, www.foodnavigator.com/Article/2019/05/20/Upcycling-coffee-cherries-for-food-Every-major-chocolate-company-is-looking-at-this/.

2 Thaneeya Hawiset, "Effect of One Time Coffee Fragrance Inhalation on Working Memory, Mood, and Salivary Cortisol Level in Healthy Young Volunteers: A Randomized Placebo Controlled Trial," *Integrative medicine research* 8, no. 4 (2019), https://doi.org/10.1016/j.imr.2019.11.007.

3 Alyssa Simpson, "The History of Grappa and Its Evolution," accessed September 26, 2025, https://pietrogallusestate.com.au/the-history-of-grappa/; Alyssa Simpson, "Grappa's Origins and History," Simon Difford, accessed September 26, 2025, www.diffordsguide.com/encyclopedia/979/bws/grappas-origins-and-history.
4 "Not Just for Drinking: Winery Waste Finds New Uses," Ken Miguel, accessed September 26, 2025, https://abc7news.com/post/not-just-for-drinking-winery-waste-finds-new-uses/1977753/.
5 "National Land Cover Database 2019 (NLCD2019) Statistics for 2016," MRLC (Multi-Resolution Land Characteristics) Consortium., accessed September 26, 2025, www.mrlc.gov/data/statistics/national-land-cover-database-2019-nlcd2019-statistics-2016.
6 David R. Reidmiller et al., "Impacts, Risks, and Adaptation in the United States: The Fourth National Climate Assessment, Volume II" (2018).
7 "Climate Change Indicators: Wildfires," U.S. Environmental Protection Agency, accessed September 26, 2025, www.epa.gov/climate-indicators/climate-change-indicators-wildfires.

References

Difford, Simon. "Grappa's Origins and History." Accessed September 26, 2025. www.diffordsguide.com/encyclopedia/979/bws/grappas-origins-and-history.

Hawiset, Thaneeya. "Effect of One Time Coffee Fragrance Inhalation on Working Memory, Mood, and Salivary Cortisol Level in Healthy Young Volunteers: A Randomized Placebo Controlled Trial." *Integrative Medicine Research* 8, no. 4 (2019): 273–78. https://doi.org/10.1016/j.imr.2019.11.007.

Miguel, Ken. "Not Just for Drinking: Winery Waste Finds New Uses." 2017. Accessed September 26, 2025. https://abc7news.com/post/not-just-for-drinking-winery-waste-finds-new-uses/1977753/.

MRLC (Multi-Resolution Land Characteristics) Consortium. "National Land Cover Database 2019 (NLCD2019) Statistics for 2016." Accessed September 26, 2025. www.mrlc.gov/data/statistics/national-land-cover-database-2019-nlcd2019-statistics-2016.

Reidmiller, David R., Christopher W. Avery, and David R. Easterling, Kenneth E. Kunkel, K. L. M. Lewis, Thomas K. Maycock, and Brooke C. Stewart, eds. *Impacts, Risks, and Adaptation in the United States: The Fourth National Climate Assessment*, Volume II. Washington, DC: U.S. Global Change Research Program, 2018. https://doi.org/10.7930/NCA4.2018

Southey, Flora. "Upcycling Coffee Cherries for Food: "Every Major Chocolate Company Is Looking at This." 2019. Accessed September 26, 2025. www.foodnavigator.com/Article/2019/05/20/Upcycling-coffee-cherries-for-food-Every-major-chocolate-company-is-looking-at-this/.

Simpson, Alyssa. "The History of Grappa and Its Evolution." Pietro Gallus Estate, published ca. 2018. Accessed September 26, 2025. https://pietrogallusestate.com.au/the-history-of-grappa/.

U.S. Environmental Protection Agency. "Climate Change Indicators: Wildfires." Accessed September 26, 2025. https://19january2021snapshot.epa.gov/climate-indicators/climate-change-indicators-wildfires_.html#:~:text=This%20map%20shows%20the%20average,be%20included%20in%20this%20analysis.

PART III
Beyond

Ehsan Baharlou

Beyond Plastics discusses a transformative shift in utilizing biopolymers in architecture and design. A transferential paradigm that leverages renewable and biodegradable biopolymers with respect to the circular economic model to highlight alternative approaches that were neglected in building construction.

In *A Path Toward the Biomaterials Age*, Javier Fernandez and Stylianos Dritsas emphasize a new paradigm in manufacturing that moves away from the synthetic plastics industry toward a sustainable society through the use of biological materials and design. They envision a biomaterial age characterized by the seamless integration of ecological life cycle with human-made products. The authors showcase this manifestation through Fungus-Like Additive Materials (FLAM), which are biodegradable and cost-effective alternatives that align with circular economic models.

The biomaterial revolution is further explored in this section by introducing additional alternatives with high potential for use in manufacturing, such as chitin and its derivative chitosan. In *ChitoFold: Designing 3D Forms from 2D Gels Toward a Fuel-Free Future*, Laia Mogas-Soldevila explores a multi-scale design and fabrication methodology using water-based chitosan biopolymers, from the nano-scale to the macro-scale, by engineering polymer chains, controlling 3D print direction, and aligning stress line toolpaths to achieve the desired shapes. This innovative approach induces anisotropic self-folding, revolutionizing new aesthetics of design with biopolymers. In *Chitin-Based Biocomposites: Sustainable Alternatives for Architectural Applications*, Paulina Grabowska and Hanaa Dahy present research on chitin- and chitosan-based biocomposites reinforced with natural fibers such as flax. These natural fiber-reinforced composites (NFRCs) are proposed as an alternative to conventional composites derived from synthetic polymers and are showcased in various applications, from furniture to facade and wall elements.

The new age of biomaterials in building construction is establishing its presences through demonstrators at building scales. In *Toward Natural Fiber Architecture: From Synthetic to Bio-Based Composite Building Systems*, the authors present a paradigm shift from synthetic fibers and petrochemical polymers to renewable and bio-based polymers. They showcase a series of large-scale architectural demonstrators, such as the BUGA Fibre Pavilion and livMatS Pavilion,

DOI: 10.4324/9781003613589-13

to explore novel natural and bio-based fiber composites in architecture, aiming to reframe composite engineering in architecture toward climate-positive, circular practices in architecture.

Exploring biopolymer composites expands in *Malleability as a Driver for Design: Activating Alternative Resource Flows with 3D-Printed Biopolymer Composites*, where the authors present regenerative architecture practice by introducing the concept of "malleability." Within this framework, they embrace both material properties and design strategies to enhance sustainable features such as adaptability, responsiveness, and circularity. They further explore this concept in the project Rhizerial, a biopolymer composite vault structure, which revisits geological considerations of biopolymer history in connection with contemporary environmental concerns.

11

A PATH TOWARD THE BIOMATERIALS AGE

Javier G. Fernandez and Stylianos Dritsas

The Rise and Fall of Plastic Manufacture

Over the span of almost a century since their inception, synthetic plastics have facilitated an accelerating technological and scientific progress in areas as broad and diverse as from medicine to aeronautics, providing the basis for inexpensive manufacturing with highly controllable synthetic materials. Their influence on shaping the modern way of life was so pivotal that they often are credited as a catalyst for the emergence of a consumption-oriented society. This change of paradigm, and consequential societal transformation induced by plastics, occurred because plastics offered technological possibilities unattainable by their predecessors. Plastics are not merely cheap and easy to shape, they are also lightweight, antimicrobial, and excellent barriers for water and oxygen, and in general, they have an extraordinary capacity to withstand the effects of the natural elements. Yet exactly these key advantages are also their Achilles' heel; plastics are unique among large-scale manufactured materials because they are entirely artificial and disconnected from Earth's natural ecological cycles. Their production, as well as their recovery, depends solely on human intervention. Unfortunately, efforts to reclaim the embodied resources in plastics are often sporadic or nearly absent, which has led to most of the plastics produced historically accumulating in landfills, the oceans, and other environments that we are only beginning to understand.[1]

Unfortunately, we now are faced with a challenging circumstance: Over the past few decades, public awareness of the environmental effects of our actions has grown considerably. Today, a significant part of that concern focuses on the impact that plastics have on our ecosystems.[2] Moreover, unlike other environmental considerations, the problem of plastics is undeniably anthropogenic, and its consequences are easily observable. Nonetheless, this heightened awareness has led to only cautious regulations, largely because few viable alternatives exist for such an indispensable material that underpins our economy and modern lifestyle.[3] It is nearly unfeasible to impose strict regulations on the excessive use of plastics. Consequently, most current solutions focus on recovering plastics, primarily through recycling practices.[4] However, as the World Bank Group recently confirmed,[5] current recovery processes tied to uncontrolled plastic production are technologically unfeasible for developing countries and have led to waste management systems in developed nations that are both oversized and increasingly unaffordable.[6]

DOI: 10.4324/9781003613589-14

Our research aims to pioneer new materials and production methods that move away from the industrial legacy of synthetic polymers. Rather than relying on incremental adjustments, we see this as an opportune moment for a transformative shift in approach. We envision a world where human-made products are seamlessly integrated into their ecological life-cycles—from creation to reclamation—mirroring the processes found in nature. Achieving this goal requires developing technologies that emulate natural production and consumption patterns. Our work on additive manufacturing using Fungus-Like Adhesive Materials serves as a case study, illustrating both the challenges and possibilities that could pave the path toward the era of biomaterials.

Biomaterials as Driving Force for a New Economy

Regardless of their dominance for thousands of years starting from prehistoric uses of timber and bone, biological materials today are largely superseded by industrial material products, past the technological revolution that took place about 200 years ago. Interest in biological materials across the scientific community was reignited approximately two decades ago with the emergence of the rapidly developing field of tissue engineering.[7] Tissue engineering, the medical branch of materials science focused on replacing and enhancing organ functions, demands materials where mechanical characteristics are considered only after ensuring their suitability to interact with the human body.[8] In this context, the extraction and regeneration of biological molecules, preservation of their organizational design in decellularized tissues, and reproduction as biomimetic materials, became central to the field.[9] Tissue engineering highlighted the inability of existent manufacturing materials to interact with living components and generated a demand for new materials with biological qualities.

Conceptually, the challenges faced by tissue engineers are similar to those of sustainable manufacturing, where the difficulty is not with the design of a well-functioning solution in isolation, but one capable of interacting with its living host. Despite the obvious differences between the human body and Earth's ecosystem, interestingly, both fields are converging to the same conclusion; there are no better-integrated materials than those already belonging to the host system.[10]

Yet, despite remarkable advances in reproducing biological composites for medical applications, adopting these highly controlled biomaterials for general manufacturing is hampered by trade-offs between production complexity and cost, limiting their large-scale use.[11] The underlying obstacle to industrializing such materials is not in the nature of the materials themselves but in their lack of fit within an industry strongly conditioned by the principles of plastics. Challenges that prevent biomaterials from scaling are in their intrinsic complexity, compared to synthetic polymers, designed for ease of manufacturing; the production speed of current biofabrication technologies, which does neither allow spanning the orders of magnitude exhibited in natural hierarchical structures nor the efficiencies of mass human-made production; and the aggregate costs per unit volume of production, including both materials and tooling. Nevertheless, the advancement of digital manufacturing, in both software and hardware, that took place over merely the past few decades has already eased some of those challenges.

An illustrative example of leveraging natural biomaterials for industrial-scale production as a strategy to counteract our reliance on plastics is demonstrated by the development of the Fungus-Like Additive Materials 3D printing process.[12] Fungus-Like Additive Material (FLAM) is a reproduction of the composition and design of the cell wall of oomycetes. Oomycetes are

FIGURE 11.1 The natural composite pillar was a 5-meter-tall prototype which was assembled using FLAM out 50 segments 3D printed in 60 hours total. It weighed approximately 100 kg and its material cost was merely S$360. (Photography by Frank Pinkers).

a class of microorganisms that closely resemble fungi; however, unlike fungi, they construct their cell walls primarily from cellulose, with low amounts of another biopolymer, chitin.[13] When this composition is artificially replicated, it produces materials that are lightweight, priced comparably to the cheapest plastics, and capable of constructing large three-dimensional structures, see Figure 11.1. These attributes closely match many of the current functionalities and economic benefits of industrial plastics. FLAM's properties are in the range of high-density synthetic foams and low-density timbers. They are extruded and cured at room temperature, requiring low energy input compared to thermoplastics. The 3D printing speed for FLAM is at least one order of magnitude higher than that of conventional plastics because there is no requirement for the material to melt before being deposited.[14] Objects created by FLAM are naturally biodegradable because their constituent components, cellulose and chitin, are not chemically modified, so the biocomposite behaves like a thermoplastic.

A fascinating and often overlooked aspect of biomaterials is that they hold the potential for a fully decentralizing production of consumer goods. The building blocks of FLAM, namely cellulose and chitin, are among the most abundant and widely distributed organic materials in the Earth's crust. Moreover, they are produced and degraded in large quantities across virtually every ecosystem on the planet.[15] These characteristics open the door for production models in which materials are sourced, processed, and broken down within a closed-loop or regional system. This approach would eliminate the need for extensive component transport and develop large-scale synthesis and recovery infrastructures to match production.[16] The potential for circular manufacturing, that is perpetually producing and consuming resources within a localized ecosystem, with FLAM was demonstrated via an experiment where cellulose was sourced from waste paper and chitin was extracted from insects deployed in insect-farming for food waste bioconversion.[17] This highlighted that unlike conventional bioplastics which require agricultural land, use of food sources and specialized composting facilities, FLAM can be directly produced and consumed within urban metabolic cycles.[18] Despite the early stage of development of FLAM, the latest achievements obtained by bioinspired manufacture represent one of the first technologies that can compare with plastics in terms of versatility of properties, cost, or manufacturing capacity,[19] see Figure 11.2. This stands as the first clear evidence that the era of plastics is not the peak of human technological achievement. Instead, we are now witnessing the emergence of a new chapter: the biomaterial age.

The Road to Living Materials

Transition to the age of biomaterials does not require abandoning the use of plastics completely but rather the associated overuse which is the actual cause of the environmental problem. Today we do continue to use stone in construction despite the era that was dominated by it ended millennia ago. Plastics is an extraordinary family of materials with unique and difficult to match properties as a barrier to water or oxygen. Therefore, there are and will be many applications for which it does not make sense to abandon completely. However, with the existence of viable alternatives, generic use of plastics for non-functionally relevant applications will probably soon disappear, replaced by environmentally benign biomaterials, while synthetics may be relegated to coatings and thin films of specialized function, forms in which even nowadays there are alternatives that can be easily assimilated without recovery[20] and used in negligible amounts, enabling cost competitive production.

Perhaps even more crucial than the immediate technological changes brought on by a shift to biomaterials are the extensive social and economic consequences that come with it. Because biological materials are so abundant, they encourage an economic model built around local manufacturing centers and workforces that serve regions largely sourcing and consuming their own raw materials. This approach also lays the groundwork for crafting legislative frameworks to regulate the use of materials that fall outside the natural ecological cycles of that region,[21] which in turn will significantly reduce the pressure from ever-increasing recovery systems as massive and economically unfeasible as those currently required.

FIGURE 11.2 (a) The natural composite pillar prototype was created to demonstrate the scalability of the FLAM technology. (Photography by Frank Pinkers). (b) High resolution with sub-millimeter layer-height prototype 3D printed to demonstrate that FLAM can be used for rapid prototyping.

Manufacturing large-scale objects with biological materials, with costs and properties similar to those of plastics, offers early evidence toward the biomaterial age but it is still the beginning of the technological leap associated to a paradigm shift from materials sciences to product design. The mechanical properties, although extraordinary in many cases, may not be the primary or even the key differentiating factor of biological material structures. The main and unique characteristic is life itself. Life is often defined as the ability to multiply, grow, regenerate, and adapt to changes in the environment. Use of biological materials is not merely catering the substitution of synthetics such as a plastic polymers and high-embodied energy materials as an obviously sustainable alternative. It suggests for a profound change of paradigm which may give rise to new branches of science, engineering, technology and design based on the very concept of life. Progress in the control of DNA structures,[22] cellular encapsulation in hydrogels,[23] and synthesis of biomolecules by modified microorganisms[24] are technologies that, although developed for medical applications, have a direct correlation with the future technologies in materials science and their general incorporation into biological systems. Biological materials can integrate and establish reciprocal relationships with living entities, such as unicellular organisms, potentially resulting in materials that can grow, repair, or adapt to changes of the environment.[25]

The paradox in our current predicament is that until now we have tried to create materials capable of resisting the effects of time making them immune and incompatible with living organisms, resulting in unsustainable production systems we progressively come to realize. However, it is precisely the adaptability and continuous regeneration of life which may instead become the key to producing everlasting materials. Our mere existence is the overwhelming proof that there is no mechanism more successful for the adaptation to the volatility and challenges of this planet than life itself.

Notes

1 Tang et al., "Chitin Is Endogenously Produced in Vertebrates"; Collicutt et al., "Microplastics in Juvenile Chinook Salmon and Their Nearshore Environments on the East Coast of Vancouver Island."
2 Geyer et al., "Production, Use, and Fate of All Plastics Ever Made."
3 Ausubel and Waggoner, "Dematerialization: Variety, Caution, and Persistence."
4 Shaxson, "Structuring Policy Problems for Plastics, the Environment and Human Health: Reflections from the UK."
5 Hoornweg et al., "Environment: Waste Production Must Peak This Century."
6 Dyson and Chang, "Forecasting Municipal Solid Waste Generation in a Fast-Growing Urban Region with System Dynamics Modeling."
7 Langer and Vacanti, "Tissue Engineering."
8 Lanza et al., *Principles of Tissue Engineering*; Yang et al., "The Design of Scaffolds for Use in Tissue Engineering. Part I. Traditional Factors."
9 Gilbert et al., "Decellularization of Tissues and Organs"; Badylak et al., "Whole-Organ Tissue Engineering: Decellularization and Recellularization of Three-Dimensional Matrix Scaffolds."
10 Shin et al., "Biomimetic Materials for Tissue Engineering"; Ragauskas et al., "The Path Forward for Biofuels and Biomaterials"; Álvarez-Chávez et al., "Sustainability of Bio-Based Plastics: General Comparative Analysis and Recommendations for Improvement."
11 Fernandez and Ingber, "Manufacturing of Large-Scale Functional Objects Using Biodegradable Chitosan Bioplastic"; Vincent, "The Trade-off: A Central Concept for Biomimetics."
12 Sanandiya et al., "Large-Scale Additive Manufacturing with Bioinspired Cellulosic Materials."
13 Lévesque, "Fifty Years of Oomycetes—from Consolidation to Evolutionary and Genomic Exploration."
14 Dritsas et al., "An Additive and Subtractive Process for Manufacturing with Natural Composites."
15 Cohen, "Chitin Biochemistry: Synthesis and Inhibition"; Klemm et al., "Cellulose: Fascinating Biopolymer and Sustainable Raw Material."
16 Stahel, "The Circular Economy."
17 Sanandiya et al., "Circular Manufacturing of Chitinous Bio-Composites via Bioconversion of Urban Refuse."
18 Kennedy et al., "The Changing Metabolism of Cities."
19 Sanandiya et al., "Large-Scale Additive Manufacturing with Bioinspired Cellulosic Materials"; Kumar et al., "Characterization of the Electrical, Acoustic, and Thermal Insulating Properties of Biodegradable Chito-Cellulosic Structural Composites."
20 Kale et al., "Compostability of Bioplastic Packaging Materials: An Overview."
21 Convery et al., "The Most Popular Tax in Europe? Lessons from the Irish Plastic Bags Levy"; Krass et al., "Environmental Taxes and the Choice of Green Technology."
22 Rothemund, "Folding DNA to Create Nanoscale Shapes and Patterns."
23 Fernandez and Khademhosseini, "Micro-Masonry: Construction of 3D Structures by Microscale Self-Assembly."
24 Nguyen et al., "Engineered Living Materials: Prospects and Challenges for Using Biological Systems to Direct the Assembly of Smart Materials."
25 Van Tittelboom et al., "Use of Bacteria to Repair Cracks in Concrete."

References

Álvarez-Chávez, Clara Rosalía, Sally Edwards, Rafael Moure-Eraso, and Kenneth Geiser. "Sustainability of Bio-Based Plastics: General Comparative Analysis and Recommendations for Improvement." *Journal of Cleaner Production* 23, no. 1 (2012): 47–56. https://doi.org/10.1016/j.jclepro.2011.10.003.

Ausubel, Jesse H., and Paul E. Waggoner. "Dematerialization: Variety, Caution, and Persistence." *Proceedings of the National Academy of Sciences* 105, no. 35 (2008): 12774–79. https://doi.org/10.1073/pnas.0806099105.

Badylak, Stephen F., Doris Taylor, and Korkut Uygun. "Whole-Organ Tissue Engineering: Decellularization and Recellularization of Three-Dimensional Matrix Scaffolds." *Annual Review of Biomedical Engineering* 13, no. 1 (2011): 27–53. https://doi.org/10.1146/annurev-bioeng-071910-124743.

Cohen, E. "Chitin Biochemistry: Synthesis and Inhibition." *Annual Review of Entomology* 32, no. 1 (1987): 71–93. https://doi.org/10.1146/annurev.en.32.010187.000443.

Collicutt, Brenna, Francis Juanes, and Sarah E. Dudas. "Microplastics in Juvenile Chinook Salmon and Their Nearshore Environments on the East Coast of Vancouver Island." *Environmental Pollution* 244 (January 2019): 135–42. https://doi.org/10.1016/j.envpol.2018.09.137.

Convery, Frank, Simon McDonnell, and Susana Ferreira. "The Most Popular Tax in Europe? Lessons from the Irish Plastic Bags Levy." *Environmental and Resource Economics* 38, no. 1 (2007): 1–11. https://doi.org/10.1007/s10640-006-9059-2.

Dritsas, Stylianos, Yadunund Vijay, Marina Dimopoulou, Naresh Sanadiya, and Javier G. Fernandez. "An Additive and Subtractive Process for Manufacturing with Natural Composites." In *Robotic Fabrication in Architecture, Art and Design 2018*. Springer International Publishing, 2018. https://doi.org/10.1007/978-3-319-92294-2_14.

Dyson, Brian, and Ni-Bin Chang. "Forecasting Municipal Solid Waste Generation in a Fast-Growing Urban Region with System Dynamics Modeling." *Waste Management* 25, no. 7 (2005): 669–79. https://doi.org/10.1016/j.wasman.2004.10.005.

Fernandez, Javier G., and Donald E. Ingber. "Manufacturing of Large-Scale Functional Objects Using Biodegradable Chitosan Bioplastic." *Macromolecular Materials and Engineering* 299, no. 8 (2014): 932–38. https://doi.org/10.1002/mame.201300426.

Fernandez, Javier G., and Ali Khademhosseini. "Micro-Masonry: Construction of 3D Structures by Microscale Self-Assembly." *Advanced Materials* 22, no. 23 (2010): 2538–41. https://doi.org/10.1002/adma.200903893.

Geyer, Roland, Jenna R. Jambeck, and Kara Lavender Law. "Production, Use, and Fate of All Plastics Ever Made." *Science Advances* 3, no. 7 (2017). https://doi.org/10.1126/sciadv.1700782.

Gilbert, Thomas W., Tiffany L. Sellaro, and Stephen Badylak. "Decellularization of Tissues and Organs." *Biomaterials*, 27, no. 19 (March 2006): 3675–3683. https://doi.org/10.1016/j.biomaterials.2006.02.014.

Hoornweg, Daniel, Perinaz Bhada-Tata, and Chris Kennedy. "Environment: Waste Production Must Peak This Century." *Nature* 502, no. 7473 (2013): 615–17. https://doi.org/10.1038/502615a.

Kale, Gaurav, Thitisilp Kijchavengkul, Rafael Auras, Maria Rubino, Susan E. Selke, and Sher Paul Singh. "Compostability of Bioplastic Packaging Materials: An Overview." *Macromolecular Bioscience* 7, no. 3 (2007): 255–77. https://doi.org/10.1002/mabi.200600168.

Kennedy, Christopher, John Cuddihy, and Joshua Engel-Yan. "The Changing Metabolism of Cities." *Journal of Industrial Ecology* 11, no. 2 (2007): 43–59. https://doi.org/10.1162/jie.2007.1107.

Klemm, Dieter, Brigitte Heublein, Hans-Peter Fink, and Andreas Bohn. "Cellulose: Fascinating Biopolymer and Sustainable Raw Material." *Angewandte Chemie International Edition* 44, no. 22 (2005): 3358–93. https://doi.org/10.1002/anie.200460587.

Krass, Dmitry, Timur Nedorezov, and Anton Ovchinnikov. "Environmental Taxes and the Choice of Green Technology." *Production and Operations Management* 22, no. 5 (2013): 1035–55. https://doi.org/10.1111/poms.12023.

Kumar, Ajay P.V., Shiwei Ng, Yung Boon Chong, Heow Pueh Lee, Stylianos Dritsas, and Javier G. Fernandez. "Characterization of the Electrical, Acoustic, and Thermal Insulating Properties of

Biodegradable Chito-Cellulosic Structural Composites." *Sustainable Materials and Technologies* 44 (2025): e01394. https://doi.org/10.1016/j.susmat.2025.e01394.

Langer, Robert, and Joseph P. Vacanti. "Tissue Engineering." *Science* 260, no. 5110 (1993): 920–26. https://doi.org/10.1126/science.8493529.

Lanza, Robert, Robert Langer, and Joseph Vacanti, eds. *Principles of Tissue Engineering*. Elsevier, 2014. https://doi.org/10.1016/c2011-0-07193-4.

Lévesque, C. André. "Fifty Years of Oomycetes—From Consolidation to Evolutionary and Genomic Exploration." *Fungal Diversity* 50, no. 1 (2011): 35–46. https://doi.org/10.1007/s13225-011-0128-7.

Nguyen, Peter Q., Noémie-Manuelle Dorval Courchesne, Anna Duraj-Thatte, Pichet Praveschotinunt, and Neel S. Joshi. "Engineered Living Materials: Prospects and Challenges for Using Biological Systems to Direct the Assembly of Smart Materials." *Advanced Materials* 30, no. 19 (2018). https://doi.org/10.1002/adma.201704847.

Ragauskas, Arthur J., Charlotte K. Williams, Brian H. Davison, et al. "The Path Forward for Biofuels and Biomaterials." *Science* 311, no. 5760 (2006): 484–89. https://doi.org/10.1126/science.1114736.

Rothemund, Paul W. K. "Folding DNA to Create Nanoscale Shapes and Patterns." *Nature* 440, no. 7082 (2006): 297–302. https://doi.org/10.1038/nature04586.

Sanandiya, Naresh D., Christoph Ottenheim, Jun Wei Phua, Augusta Caligiani, Stylianos Dritsas, and Javier G. Fernandez. "Circular Manufacturing of Chitinous Bio-Composites via Bioconversion of Urban Refuse." *Scientific Reports* 10, no. 1 (2020). https://doi.org/10.1038/s41598-020-61664-1.

Sanandiya, Naresh D., Yadunund Vijay, Marina Dimopoulou, Stylianos Dritsas, and Javier G. Fernandez. "Large-Scale Additive Manufacturing with Bioinspired Cellulosic Materials." *Scientific Reports* 8, no. 1 (2018). https://doi.org/10.1038/s41598-018-26985-2.

Shaxson, Louise. "Structuring Policy Problems for Plastics, the Environment and Human Health: Reflections from the UK." *Philosophical Transactions of the Royal Society B: Biological Sciences* 364, no. 1526 (2009): 2141–51. https://doi.org/10.1098/rstb.2008.0283.

Shin, Heungsoo, Seongbong Jo, and Antonios G. Mikos. "Biomimetic Materials for Tissue Engineering." *Biomaterials* 24, no. 24 (2003): 4353–64. https://doi.org/10.1016/s0142-9612(03)00339-9.

Stahel, Walter R. "The Circular Economy." *Nature* 531, no. 7595 (2016): 435–38. https://doi.org/10.1038/531435a.

Tang, W. Joyce, Javier G. Fernandez, Joel J. Sohn, and Chris T. Amemiya. "Chitin Is Endogenously Produced in Vertebrates." *Current Biology* 25, no. 7 (2015): 897–900. https://doi.org/10.1016/j.cub.2015.01.058.

Van Tittelboom, Kim, Nele De Belie, Willem De Muynck, and Willy Verstraete. "Use of Bacteria to Repair Cracks in Concrete." *Cement and Concrete Research* 40, no. 1 (2010): 157–66. https://doi.org/10.1016/j.cemconres.2009.08.025.

Vincent, Julian F.V. "The Trade-off: A Central Concept for Biomimetics." *Bioinspired, Biomimetic and Nanobiomaterials* 6, no. 2 (2017): 67–76. https://doi.org/10.1680/jbibn.16.00005.

Yang, Shoufeng, Kah-Fai Leong, Zhaohui Du, and Chee-Kai Chua. "The Design of Scaffolds for Use in Tissue Engineering. Part I. Traditional Factors." *Tissue Engineering* 7, no. 6 (2001): 679–89. https://doi.org/10.1089/107632701753337645.

12

CHITOFOLD

Designing 3D Forms from 2D Gels Toward a Fuel-Free Future

Laia Mogas-Soldevila

Introduction to Bio-Based Polymers

Advances in materials science, along with the growing availability of protocols and open-source material databases, have significantly changed how bio-based polymer composites are approached. Thanks to their safe processing properties, bio-composites are now being explored outside of specialized labs. Coupled with new digital design and fabrication techniques, this shift enables designers to experiment by making unique properties emerge and creating new functions that contribute to both human and planetary health.

Of particular interest to DumoLab's work are structural biomaterials.[1] These are naturally occurring polymers that provide mechanical strength and durability while still biodegrading at end of life, which is desired in architectural applications that aim at avoiding unsustainable binders from fuel-based plastics. DumoLab Research (DLR),[2] at the Penn Stuart Weitzman School of Design, is an interdisciplinary laboratory that specializes in the translation of biomaterial systems based on structural biomaterials from biomedical engineering, life sciences, and vernacular architecture into everyday products and structures that support health during processing, use, and end of life. Research is focused on ambient-conditions additive manufacturing, water-based materials processing, bioremediation with living building materials, and bio-based alternatives to unsustainable traditional materials. The work at DumoLab, continuing the multi-year and multi-institution research collaboration described in this chapter, re-engineers these structural biopolymers to bind natural aggregates and fillers into much-needed hyper-sustainable bio-composite structures while considering their durability and mechanical demands in products and building parts, their manufacturability with robotic fabrication systems, their economic viability, and their aesthetic value toward widespread adoption.

In that sense, biomaterial building blocks widely used in scientific fields, are reconsidered for design applications. For instance, silk fibroin is a protein found in silk produced by silkworms and other insects with many uses in biotechnology and biomedical engineering beyond traditional textiles. Collagen is a protein that is a major structural component of skin, muscles, bones, tendons, and other connective tissues. Both silk fibroin and collagen have been used in fashion and upholstery leather replacements in recent bio-design work.[3] Keratin is a

DOI: 10.4324/9781003613589-15

biopolymer with a high cysteine content that allows for the formation of disulfide bonds making strong and resilient tissue such as hair and nails. For instance, human hair waste has potential to enhance construction materials as a fibrous filler and contribute to pollution control by, for example, absorbing spills.[4] Lignin is a complex aromatic polymer found in plant cell walls that provides structural support and rigidity, and importantly, is highly resistant to degradation. It has been recently used in bio-composite canopies from food waste to confer stiffness.[5] Starch is a polysaccharide found in plants that helps nurture and store energy in animals and can be used to thicken foods as well as bio-composite inks for printing. Cellulose, present in plants and trees, is the most abundant biopolymer, a biodegradable linear polymer made of glucose units that forms hydrogen bonds. It can be used in powder or fiber form in bio-composites and helps with tensile performance at the macroscale.[6] Chitin, from fungi, insect, and crustacean shells, is the second most abundant biopolymer on Earth with a molecular structure that allows for strong hydrogen bonding. It can be processed into chitosan which is soluble in acidic water and can be printed into plastic-like structures as performed in ChitoFold, the project discussed in this text[7] which was developed over a decade by an interdisciplinary team from MIT, Harvard, Penn, and AMOLF.

Recent research is bringing biomaterial blends—typically used in small-scale medical applications like implants, resorbable electronics, tissue scaffolding, and drug delivery systems—into the realm of large-scale digital design and fabrication. These biomaterials can be processed in a non-toxic, bottom-up way, using abundant and easily bio-synthesized resources. Their composites can be both technological—offering mechanical strength, microstructural performance, and robustness—and biological—being living, adaptable, and bio-functional—making them ideal for applications at the interface of these two domains.[8] The sustainable nature of biomaterials during production and after use means that projects using them can minimize their ecological impact, relying on low-energy, non-toxic, zero-waste design and fabrication methods.

With computational design and hybrid fabrication methods such as 2.5D printing, inkjet printing, screen-printing, and specialized molding, biomaterial blends can be transformed into custom objects and structures. The results often lead to surprising, innovative outcomes. ChitoFold harnesses 2.5D-printing, an additive manufacturing technique that enables hierarchical distribution of low viscosity inks—such as water-based biomaterial blends—on a flat print bed, and, in the case of ChitoFold, it is based on the first large-scale platform of this kind,[9] and ensures further emergence of 3D shape by the very nature of the hierarchies described in toolpaths (Figure 12.1).

Design Principles in Responsive Biomaterials

ChitoFold highlights digitally created constructs that react to their environment by geometry and humidity-responsive folding, but further functionalization of any biomaterial described above can lead to, for instance, sensing and displaying factors like health, performance, mood, stress, and temperature,[10] or even to include unique features in everyday objects like anisotropic mechanics, self-repairing capabilities, color-changing properties, or even the release of therapeutic substances, scents, or nutrients.[11] These properties allow for a range of multifunctional design applications that balance both high performance and aesthetic appeal. They are diverse and can impact everyday products, from apparel and footwear to fashion accessories, furniture, and building components.

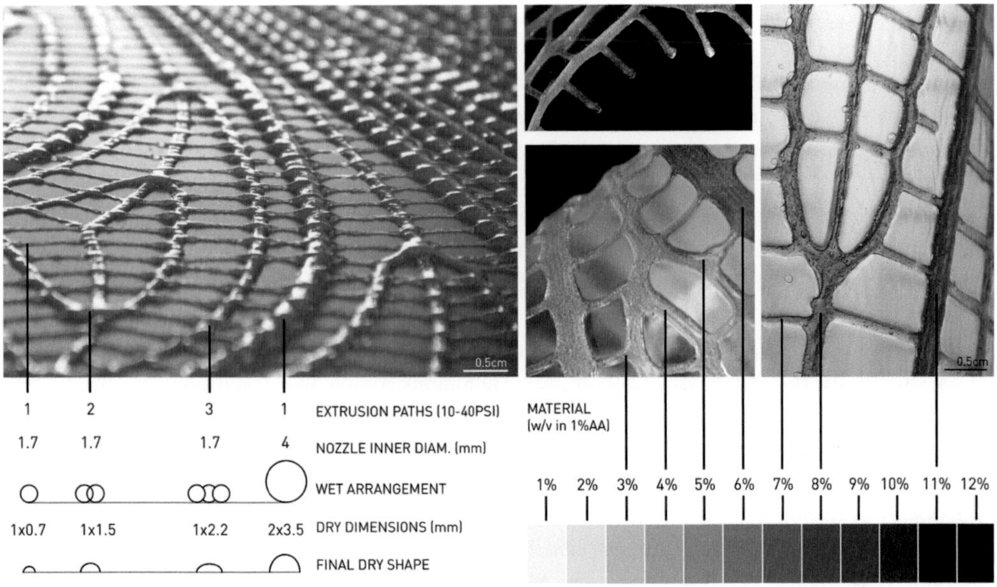

FIGURE 12.1 An example of 2.5D-printing of water-based structural biopolymers demonstrating material distribution patterns in cross section and resulting dry constructs. Different concentrations of gels (in this case chitosan) generate a gradient of materials with different opacities, viscosities, and stiffnesses. Credits: Jorge Duro-Royo, Laia Mogas-Soldevila, and Neri Oxman (MIT Mediated Matter Group, 2015).

Recently developed advanced materials are capable of morphing in response to various stimuli inspired by biological systems. Scientists are increasingly interested in creating materials that can adapt their shape or structure the way biological organisms naturally change form in response to their environment.

In biological systems, shape changes often occur as a response to environmental factors, through the integrated interaction of material, structure, and form. These changes are typically driven by internal forces within the system, rather than external mechanical pressures. Biological systems use pre-programmed strategies to adjust their shape. This includes the controlled alignment of material reinforcement such as fibrous regions in plants and trees. It also manifests in uneven distribution of material reinforcement like in suture joints in skulls, and can also be appreciated in the breakdown and rearrangement of material structures in self-healing process across skins. The first two strategies often involve materials that respond to the environment without requiring cellular energy input, making them ideal for research into programmable materials and active composites. *At the nanoscale*, for example, DNA naturally assembles into helical structures through a process called chiral twisting, while polypeptide chains fold into 3D proteins with various properties and functions. On the microscale, plants like climbing vines and orchid seed pods shape themselves passively by seeking energy-efficient configurations driven by reinforcement in their internal fiber structures. Water is also a key factor in plant self-shaping, where changes in water content cause tissues in leaves, stems, and roots to fold or swell, based on internal pressures. These processes combine at various levels, from microscale geometric patterns to nanoscale chemistry, to drive tissue deformation. Similarly, without using metabolic energy, pinecones can open when their water content changes, driven by programmed responses to moisture. *At the meso-scale*, certain worms change shape passively by rearranging their fibrous membranes, and sea cucumbers can rapidly adjust the stiffness of their collagen fibers in response to chemical cues. *At the macro-scale*, trees adjust the angle of cellulose fibers in response to environmental factors like gravity, wind, or damage, enabling them to bend and bear loads more effectively.

Drawing inspiration from the mentioned natural strategies, research into synthetic materials that can self-fold or change shape in a controlled way is opening a wide array of potential applications, from small-scale medical devices to large-scale components like those in aircraft.[12] Current techniques for inducing folding in materials can be categorized by their scale and intended use. *At the nanoscale*, "DNA origami" allows for the precise spatial arrangement of DNS strands by folding them into complex 3D shapes, much like traditional paper origami but on a molecular level. This allows the creation of nanostructures able to carry drugs or to specific locations in the body to perform highly specialized tasks such as cancer therapy. *At the microscale*, smart materials, such as thin polymer films, can be engineered to fold into 3D shapes in response to specific stimuli like light, heat, electric fields, or chemical changes. Bio-inspired design principles are applied in these processes to manipulate internal stresses, enabling controlled shape transformation. Methods include folding bilayer materials, adjusting 3D shapes from non-equilibrium states, and using advanced techniques like iono-printing and programmable 3D hydrogels. *At the mesoscale*, functional robots are being developed that fold from 2D flat sheets into 3D forms using techniques like Joule heating or mechanisms like pop-up books. In addition, laser-engraved origami with moving parts or spring elements has been used for creating meso-scale self-assembling structures. Emerging "4D printing" technologies, which enable materials to change shape over time, are also being used to print multi-material composites that respond to environmental triggers.[13] These technologies often rely on composite inks that exhibit specific elastic and swelling behaviors and use dynamic materials like shape-memory fibers or hydrophilic hinges within flexible matrices. For example, 3D printing inspired by balsa wood enables the creation of lightweight, cellular composites reinforced with high-aspect-ratio fibers, which are oriented as the material flows through the printer nozzle. *At the macro scale*, materials like wood—known for its moisture sensitivity—are being used to design self-folding architectural components.[14] These components can change shape based on changes in humidity, with considerable dimensional shifts across the grain, ultimately displaying large building-integrated assemblies that can passively control indoor light and ventilation. Using digitally designed and fabricated fiber-reinforced polymers (FRP), kinematic systems can be bioinspired from plant motion mechanics.[15] Elastic deformation of entire structures replaces the need for local hinges enabled by designed venations combining high tensile strength with low bending stiffness, enabling large scale deployable structures. Additionally, large-scale shape-changing techniques have been applied to aircraft wings, using elastomeric materials that adapt to environmental changes during flight. This capability improves flight efficiency at various altitudes, reduces the weight of the structure, boosts fuel economy, and minimizes environmental and noise impacts.

Designing Anisotropy in Large-Scale Printed Biopolymers

If controlled at the macroscale, these self-folding and morphing materials have numerous applications. For example, they could be used in architecture to create adaptive structures that change shape in response to environmental conditions, or in aerospace to produce wings that adjust during flight for improved efficiency.

Toward this goal, ChitoFold[16] describes a digital design and fabrication methodology and a predictive material simulation model, using water-based biological materials, and inducing 2D-to-3D non-reversible anisotropic self-folding within single-material manufactured objects across multiple length scales. The manufacturing system can operate at the nanoscale (via molecular chain alignment and the induction of crystallinity), micro-scale (via directional extrusion design), meso-scale (via global stress line toolpaths), and macro-scale (via architectural design), demonstrating high degrees of opportunity to tune and control both material properties and global geometric shape.

In the study, we utilized our innovative Water-Based Digital Fabrication Platform (WDFP)[17] to create large-scale hierarchical structures inspired by natural systems. This approach involves the deposition of biological materials like chitosan, a chemically modified form of chitin, along specific structural paths that reflect the primary load-bearing stress lines. The deposition can occur in various directions—longitudinal, transverse, or oblique—allowing for the integration of diverse macro-scale structural patterns into the final design. A key observation during this process is that when fully cured, the resulting structures often display unexpected, irreversible 2D-to-3D curling behaviors. One of the main objectives of this research was to explore the mechanisms that lead to the formation of these 3D geometries from simple 2D printing paths. To address this, we took a multifaceted approach: We first examined the impact of concentration-dependent crystallization and the orientation of polymer chains within the structural paths at the nanoscale using synchrotron X-ray scattering, followed by the development of a finite element model to simulate the effect of structural anisotropy on directional folding.

Design and Analysis of Anisotropic Behavior

Figure 12.2 illustrates the bio-inspired design, fabrication, and final cured chitosan-based self-folded column, which was inspired by dragonfly wings and leaf venation patterns. The WDFP employs a highly precise 6-axis robotic arm with a multi-nozzle pressure-based extruder. The extruder dispenses various concentrations of chitosan hydrogel. The system synchronizes robotic positioning and nozzle pressure to control the geometry of the final structures. These parameters are optimized for each material, ensuring consistent extrusion geometries based on material viscosity. At room temperature, the structure cures through water evaporation, resulting in a large-scale, hierarchically organized column self-folded into its final 3D form within minutes of release from print bed and remaining in that state.

Synchrotron diffraction measurements on printed and cast chitosan biopolymer samples (a technique that uses high-intensity X-rays to study the atomic structure of crystalline materials) revealed a clear correlation between chitosan concentration and crystallinity. The higher strength of dry chitosan samples was directly proportional to their degree of water-based concentration since crystallinity refers to the degree of structural order in the polymer chains. This interestingly also maps gel "color" to structural performance as we can see in the color gradient in Figure 12.1. In ChitoFold, the fact that crystallinity increases with concentration suggests that this could be used to create macroscopic samples with engineered structural anisotropy by printing. To test this, diffraction patterns were analyzed from different regions of a printed rib structure and showed that the crystallites' preferred orientation aligned with the

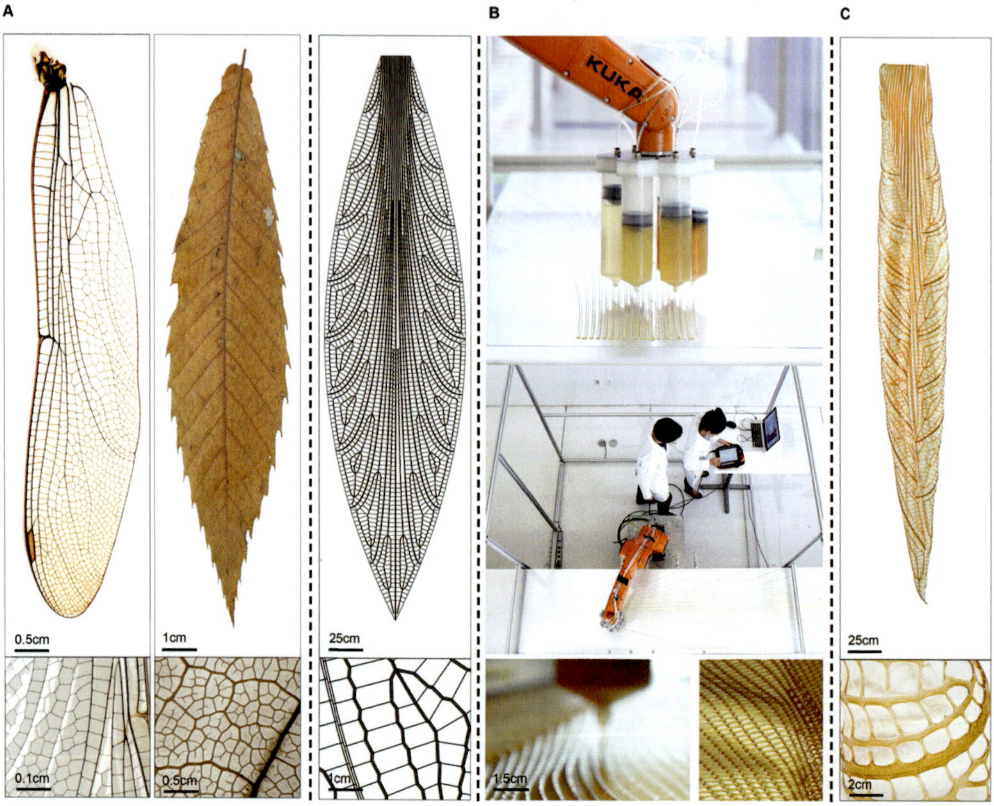

FIGURE 12.2 Overview of the Water-Based Digital Fabrication Platform (WDFP) and specific examples of manufacturable geometries (a) INSPIRATION Biomimetic case study digital design process for 2.5D toolpath generation based on dragonfly wing and leaf venation patterns of a 3 cm-long dragonfly wing specimen and a 10 cm-long dried leaf specimen. (a-right) VIRTUAL 3 m-long digitally generated toolpath with hierarchical geometric patterning. (b) PHYSICAL Fabrication platform features: robotic arm positioning system with a 1 m reach and 10 kg payload and multi-nozzle pressure deposition system carrying diverse chitosan hydrogel gel concentrations. (c) RESULTS 3 m-long self-folded column cured at ambient conditions, and a magnified view of structural folding along toolpaths induced through the fabrication process. Credits: Laia Mogas-Soldevila, Jorge Duro-Royo, Daniel Lizardo, James C. Weaver, and Neri Oxman (MIT 2017 & Penn 2024).

printing direction, further supporting the idea that material crystallinity and orientation could influence structural behavior.

To understand the factors driving the deformations in our printed constructs, we created a finite element model to simulate the effects of various printing parameters. The model incorporated real-time drying behavior, where chitosan dries faster on the top surface exposed to air than on the underside in contact with the aluminum substrate. This differential drying behavior creates a bilayer-like structure, with a rigid, fully cured top layer and a more hydrated, flexible underlayer. After removing the structure from the substrate, the underlayer shrinks, inducing curling (Figure 12.3).

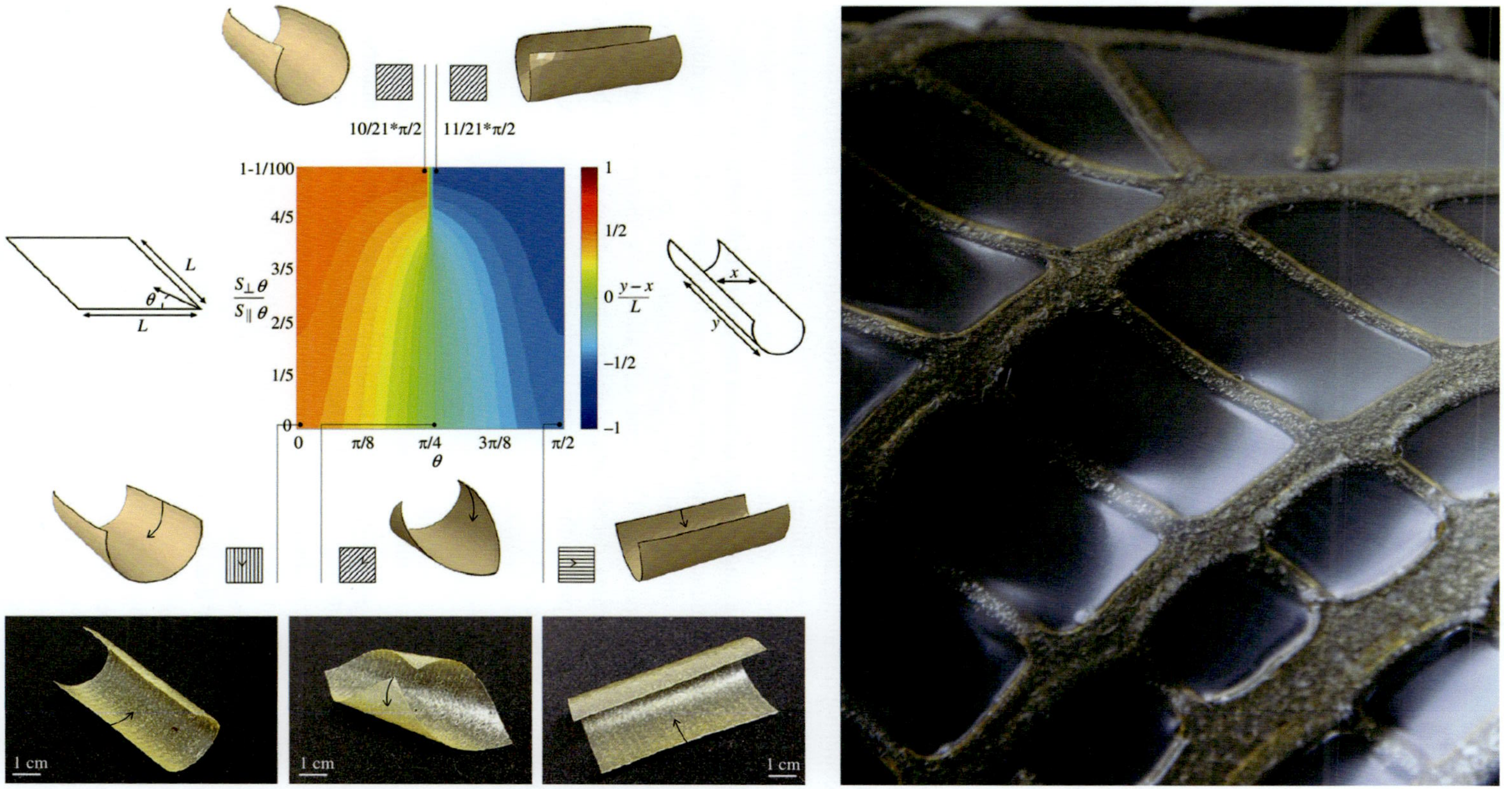

FIGURE 12.3 Numerical simulations and experiments of fabrication-induced folding as a function of structural anisotropy. Top: contour plot showing the dependency of the folding behavior on the direction of anisotropy, and the size of anisotropy introduced by varying the swelling ratio parallel and orthogonal. The folding direction is characterized by the absolute change in distance between the outer points lying on the short axis of the square. Bottom: experimental verification of the numerical model, showing a relatively high level of anisotropy, as evidenced by the direct link between print direction and folding direction (noted with arrows on two- and three-dimensional schematics and on photographs of shaped constructs) Credits: Johannes Overvelde, Laia Mogas-Soldevila, Jorge Duro-Royo, Daniel Lizardo, James C. Weaver, and Neri Oxman (MIT 2017 & Penn 2024).

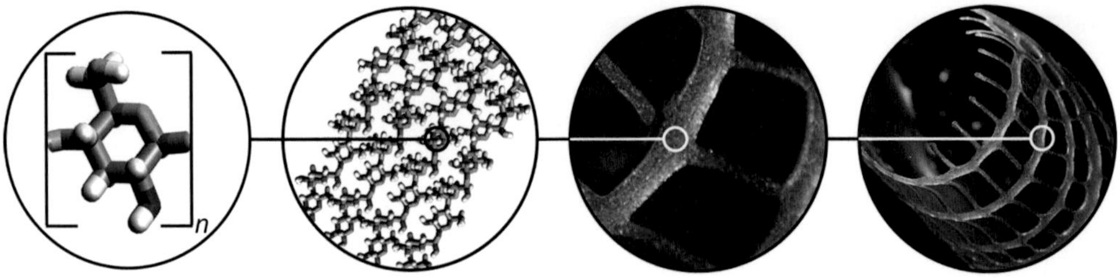

FIGURE 12.4 Schematic of the multi-scale behavior model of our material system and fabrication strategy, where the induced orientation of nano-scale chitosan crystallites ultimately drives macro-scale folding of the final additively manufactured form. (a) Chitosan monomer, (b) acetic acid-solubilized chitosan crystallites depicted following a structural rib orientation printed with the WDFP, (c) 2.5D-printed structural rib and (d) chitosan construct folded along the print direction. Credits: Laia Mogas-Soldevila, Jorge Duro-Royo, Daniel Lizardo, James C. Weaver, and Neri Oxman (MIT 2017 & Penn 2024).

Our simulations, modeled as bilayer structures showed that the direction of folding was influenced by the printing direction. For square samples, we found that when the shrinkage ratio was sufficiently high, folding occurred predictably along the print direction, but for low anisotropy, folding could happen along any direction with little control. This was demonstrated by comparing our simulations with experimental results: Printed samples folded along their diagonals, a behavior not seen in cast chitosan films, which exhibited random edge folding, supporting the conclusion that the printing process induced the anisotropic behavior.

These findings suggest that the induced nano-scale ordering from residual stresses during material extrusion, combined with the specific tool-path design, can be used to control macroscopic folding patterns (Figure 12.4). This approach allows for the programming of material responses across different scales, decoupling geometric design from material behavior. Therefore, by strategically designing print paths and material properties, we can achieve specific, predictable folding behaviors at large scales.

Toward Fuel-Free Plastic Futures

In ChitoFold, large, biodegradable structures can be 2.5D-printed using a single biopolymer, chitosan, in an aqueous form. By adjusting the concentration of chitosan, the material's crystallinity and folding behavior can be controlled. The direction of the print influences the crystallographic orientation, which can then drive folding in unexpected directions within the printed structure. This approach introduces structural anisotropy, which helps create complex, multi-scale folding patterns like the way veins form in leaves. The technology lays the foundation for designing versatile architectures that can respond to environmental stimuli with tunable folding behaviors at different scales.

Compared to other manufacturing methods that induce material anisotropy—like using magnets or fibrous fillers—this method could be scaled up more easily and affordably, especially when using simple materials and working in challenging environments. Looking

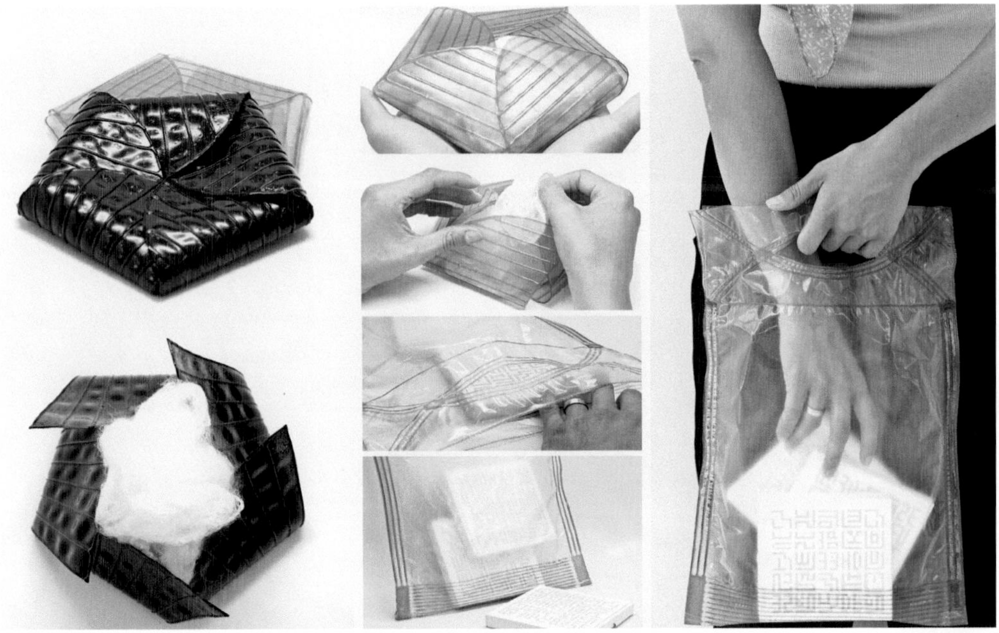

FIGURE 12.5 Packaging systems can be developed from digitally and chemically engineered bioplastic systems with desirable functional and aesthetic properties yet without toxic trace at end-of-life. Credits: Laia Mogas-Soldevila, Jorge Duro-Royo (MIT 2015).

ahead, improvements in computational software for predicting folding behavior could allow for more intricate designs and 3D outcomes. Using smaller nozzles in the extrusion printing process could enhance the resolution of printed structures; while experimenting with different material concentrations might create hybrid behaviors, combining flexibility and rigidity. The potential applications for this technology are broad, as it offers a simple, scalable method for creating complex shapes with a single material, without needing expensive equipment or multiple additives. To increase the stability and longevity of these structures—and of any others based on simple biopolymers—adding water-resistant coatings or applying cross-linking can help prevent degradation over time. Most importantly, work like this can transform fuel-based polymer manufacturing processes in industries ranging from construction to product design, but only if the water-based nature of biomaterials is embraced and designed for.

A series of packaging systems were developed from this research (Figure 12.5) to demonstrate short-term bioplastic consumables design with desirable functional and aesthetic properties and yet be able to disappear without toxic trace in rapid compostability at end-of-life.[18] A culminating ChitoFold outcome built only from chitosan, described the Ocean Pavilion (Mediated Matter Group, MIT Media Lab, 2017),[19] which presented variable-property, biomaterial-based components in a set of 3 to 5 m long structures (Figure 12.6) demonstrating that the interplay of chemical and geometrical control of an otherwise brittle material proposes functional futures for fuel-free plastic architecture.

FIGURE 12.6 Ocean Pavilion, a set of 3 to 5 m long chitosan 2.5D-printed structures using biomaterial templating of structural, manufacturing, and environmental performance. Credits: Laia Mogas-Soldevila, Jorge Duro-Royo, Daniel Lizardo, and Neri Oxman (MIT 2017).

Notes

1 Vincent, "Naturally New Materials."
2 "DumoLab Research (DLR) | Weitzman."
3 Mogas-Soldevila et al., "Additively Manufactured Leather-like Silk Protein Materials."
4 Gupta, "Human Hair 'Waste' and Its Utilization."
5 Cen et al., "CanoPIT."
6 Sanandiya et al., "Large-Scale Additive Manufacturing with Bioinspired Cellulosic Materials."
7 Mogas-Soldevila et al., "Driving Macro-Scale Transformations in Three-Dimensional-Printed Biopolymers through Controlled Induction of Molecular Anisotropy at the Nanoscale."
8 Fratzl and Weinkamer, "Nature's Hierarchical Materials."
9 Mogas-Soldevila, Duro-Royo, and Oxman, "Water-Based Robotic Fabrication: Large-Scale Additive Manufacturing of Functionally Graded Hydrogel Composites via Multichamber Extrusion"; Duro-Royo, Mogas-Soldevila, and Oxman, "Flow-Based Fabrication"; Mogas-Soldevila et al., "DESIGNING THE OCEAN PAVILION."
10 Matzeu et al., "Large-Scale Patterning of Reactive Surfaces for Wearable and Environmentally Deployable Sensors."
11 Mogas-Soldevila and Zolotovsky, "Designing with Printed Responsive Biomaterials."
12 R. Studart and M. Erb, "Bioinspired Materials That Self-Shape through Programmed Microstructures."
13 Sydney Gladman et al., "Biomimetic 4D Printing."
14 Wood et al., "Material Computation: 4D Timber Construction Towards Building-Scale Hygroscopic Actuated, Self-Constructing Timber Surfaces."
15 Lienhard et al., "Flectofin."
16 Mogas-Soldevila et al., "Driving Macro-Scale Transformations in Three-Dimensional-Printed Biopolymers through Controlled Induction of Molecular Anisotropy at the Nanoscale."

17 Mogas-Soldevila, Duro-Royo, and Oxman, "Water-Based Robotic Fabrication: Large-Scale Additive Manufacturing of Functionally Graded Hydrogel Composites via Multichamber Extrusion."
18 Mogas-Soldevila, "Water-Based Digital Design and Fabrication."
19 Mogas-Soldevila et al., "Designing the Ocean Pavilion."

References

Cen, Yuanyi, Abigail Weinstein, Andreina Sojo, Ji Yoon Bae, and Laia Mogas-Soldevila. "CanoPIT: Valorizing Unavoidable Fruit Waste into Printable Biomaterial Surfaces for Participatory Learning." In *Proceedings of the 112th Annual Meeting of the Association of Collegiate Schools of Architecture (ACSA)*, Vancouver, Canada, March 14–17, 2024. Washington, DC: Association of Collegiate Schools of Architecture, 2024. www.acsa-arch.org/chapter/canopit-valorizing-unavoidable-fruit-waste-into-printable-biomaterial-surfaces-for-participatory-learning/.

Weitzman School of Design, "About DumoLab Research (DLR)". University of Pennsylvania, accessed February 20, 2026. www.design.upenn.edu/dumolab/about.

Duro-Royo, Jorge, Laia Mogas-Soldevila, and Neri Oxman. "Flow-Based Fabrication: An Integrated Computational Workflow for Design and Digital Additive Manufacturing of Multifunctional Heterogeneously Structured Objects." *Computer-Aided Design* 69 (December 2015): 143–54. https://doi.org/10.1016/j.cad.2015.05.005.

Fratzl, Peter, and Richard Weinkamer. "Nature's Hierarchical Materials." *Progress in Materials Science* 52, no. 8 (November 1, 2007): 1263–1334. https://doi.org/10.1016/j.pmatsci.2007.06.001.

Gupta, Ankush. "Human Hair 'Waste' and Its Utilization: Gaps and Possibilities." *Journal of Waste Management* 2014, no. 1 (2014): 498018. https://doi.org/10.1155/2014/498018.

Lienhard, J., S. Schleicher, S. Poppinga, T. Masselter, M. Milwich, T. Speck, and J. Knippers. "Flectofin: A Hingeless Flapping Mechanism Inspired by Nature." *Bioinspiration & Biomimetics* 6, no. 4 (November 2011): 045001. https://doi.org/10.1088/1748-3182/6/4/045001.

Matzeu, Giusy, Laia Mogas-Soldevila, Wenyi Li, Arin Naidu, Trent H. Turner, Roger Gu, Patricia R. Blumeris, et al. "Large-Scale Patterning of Reactive Surfaces for Wearable and Environmentally Deployable Sensors." *Advanced Materials* 32, no. 28 (2020): 2001258. https://doi.org/10.1002/adma.202001258.

Mogas-Soldevila, Laia. "Water-Based Digital Design and Fabrication: Material, Product, and Architectural Explorations in Printing Chitosan and Its Composites." Thesis, Massachusetts Institute of Technology, 2015. https://dspace.mit.edu/handle/1721.1/101828.

Mogas-Soldevila, Laia, Jorge Duro-Royo, Daniel Lizardo, George G. Hollyer, Charles M. Settens, Jordan M. Cox, Johannes T. B. Overvelde, et al. "Driving Macro-Scale Transformations in Three-Dimensional-Printed Biopolymers through Controlled Induction of Molecular Anisotropy at the Nanoscale." *Interface Focus* 14, no. 3 (June 7, 2024): 20230077. https://doi.org/10.1098/rsfs.2023.0077.

Mogas-Soldevila, Laia, Jorge Duro-Royo, Daniel Lizardo, Markus Kayser, William Patrick, Sunanda Sharma, Steven Keating, John Klein, Chikara Inamura, and Neri Oxman. "Designing the Ocean Pavilion: Biomaterial Templating of Structural, Manufacturing, and Environmental Performance." *Proceedings of IASS Annual Symposia* 2015, no. 16 (August 20, 2015): 1–13.

Mogas-Soldevila, Laia, Jorge Duro-Royo, and Neri Oxman. "Water-Based Robotic Fabrication: Large-Scale Additive Manufacturing of Functionally Graded Hydrogel Composites via Multichamber Extrusion." *3D Printing and Additive Manufacturing* 1, no. 3 (September 2014): 141–51. https://doi.org/10.1089/3dp.2014.0014.

Mogas-Soldevila, Laia, Giusy Matzeu, Marco Lo Presti, and F. G. Omenetto. "Additively Manufactured Leather-like Silk Protein Materials." *Materials & Design* 203 (May 1, 2021): 109631. https://doi.org/10.1016/j.matdes.2021.109631.

Mogas-Soldevila, Laia, and Katia Zolotovsky. "Designing with Printed Responsive Biomaterials: A Review." *3D Printing and Additive Manufacturing*, 12, no. 2 (June 26, 2024): 155–68. https://doi.org/10.1089/3dp.2024.0004.

Sanandiya, Naresh D., Yadunund Vijay, Marina Dimopoulou, Stylianos Dritsas, and Javier G. Fernandez. "Large-Scale Additive Manufacturing with Bioinspired Cellulosic Materials." *Scientific Reports* 8, no. 1 (June 5, 2018): 8642. https://doi.org/10.1038/s41598-018-26985-2.

Studart, André R., and Randall M. Erb. "Bioinspired Materials That Self-Shape through Programmed Microstructures." *Soft Matter* 10, no. 9 (2014): 1284–94. https://doi.org/10.1039/C3SM51883C.

Sydney Gladman, A., Elisabetta A. Matsumoto, Ralph G. Nuzzo, L. Mahadevan, and Jennifer A. Lewis. "Biomimetic 4D Printing." *Nature Materials* 15, no. 4 (April 2016): 413–18. https://doi.org/10.1038/nmat4544.

Vincent, Julian. "Naturally New Materials." *Materials Today*, 1, no. 3 (1998): 3–6.

Wood, Dylan, David Correa, Oliver Krieg, and Achim Menges. "Material Computation: 4D Timber Construction Towards Building-Scale Hygroscopic Actuated, Self-Constructing Timber Surfaces." *International Journal of Architectural Computing* 14 (2016): 49–62.

13

CHITIN-BASED BIOCOMPOSITES

Sustainable Alternatives for Architectural Applications

Paulina Grabowska and Hanaa Dahy

Introduction

Biopolymers Binders for NFRPs

Climate change and ongoing resource depletion require from architects and designers new material solutions that help minimize CO2 emissions, reduce resource consumption, and limit waste.[1] Within this context, biopolymers – naturally occurring polymers produced by the cells of living organisms, such as starch and cellulose – are gaining importance due to their ecological and functional advantages. Biodegradable, derived from renewable sources, and widely available, biopolymers are being explored across a range of industries, from food packaging to fashion, furniture, and even architecture.[2] Among their many properties, several biopolymers also demonstrate strong adhesive capabilities and compatibility with natural fibers, making them potential replacements for synthetic resins in biocomposite materials.[3]

Composite materials are made by combining two or more components – typically a reinforcing fiber and a matrix – to create a material with improved mechanical and functional properties.[4] Natural fiber reinforced polymers (NFRPs), also referred to as biocomposites, combine natural fibers such as flax or hemp with a polymer matrix, resulting in lightweight and strong material, suitable for high-performance applications. They present a more sustainable option compared to conventional fiber-reinforced polymers (FRPs), which rely on synthetic fibers such as carbon or glass fibers. Due to these advantages, NFRPs are increasingly being used in architecture and design applications such as furniture, panels, and lightweight structural elements.[5] However, natural fibers do not ensure that the composite is fully biodegradable or recyclable. The environmental impact of NFRPs at the end of their life cycle is largely determined by the second component, the matrix. Currently, many biocomposites still rely on synthetic polymers such as polypropylene (PP), polyethylene (PE), epoxy, and polyester[6] as binders, preventing their full biodegradability and complicating their recycling. Replacing these synthetic polymers with biopolymer-based matrices opens up possibilities for fully compostable green composites.

DOI: 10.4324/9781003613589-16

This chapter focuses on chitin, a natural biopolymer extracted for example from exoskeletons of crustaceans and insects, as an alternative biodegradable binder in natural fiber reinforced composites, exploring its potential for sustainable material solutions in design and construction.

Chitin and Chitosan

Chitin, the second most abundant biopolymer in nature, is a main component of the exoskeletons of crustaceans and insects, the shells of molluscs, and the cell walls of fungi.[7] It can be extracted chemically or biologically, using renewable and circular sources, like seafood waste or mycelium.[8] Chitin's mechanical strength provides in nature a protective barrier, shielding organisms from environmental damage or predators.[9] Its hydrophobic properties enhance durability and resistance to environmental conditions. Moreover, chitin is fully biodegradable and biocompatible.[10] Combination of these properties makes chitin not only a remarkable material in nature, but also a promising ingredient for sustainable applications such as eco-friendly building materials.

Chitosan, derived from the deacetylation of chitin, retains many of its key properties, including biocompatibility and biodegradability. Unlike chitin, however, chitosan is soluble in mildly acidic solutions ($pH < 6$), which makes it easier to process into films or gels suitable for biocomposite production.[11] In the presented projects, chitosan is dissolved in acetic acid, creating gels of varying concentrations, which serve as the foundation for further material formulations.

Chitosan Flax Composites

Chitosan, having a similar structure to cellulose, allows the two biopolymers to blend and complement each other.[12] Recent studies have demonstrated the feasibility of using chitosan as a binder for cellulose-rich fibers by combining it with wood flour and cellulose, resulting in strong, durable composite material.[13] Building upon this foundation, this chapter investigates the use of chitosan gels as alternatives to conventional thermoset and thermoplastic polymers. By combining chitosan gels with locally sourced flax fibers, which are known for their good mechanical properties and high cellulose content,[14] this study aims to develop strong and fully biodegradable composite materials.

Aim

The research investigates the application of chitin-based fluids as an alternative binding material in NFRPs, focusing on their integration into established composite fabrication methods and their suitability for architectural use. Two digital fabrication techniques – Tailored Fiber Placement (TFP) and 3D printing – are employed to demonstrate the material's potential in sustainable and innovative design. Through a series of case studies, this chapter presents experimental outcomes from the BioMat research team, demonstrating the use of chitin-based biocomposites in furniture, facade systems, and wall components. Aligned with the philosophy of *Materials as a Design Tool*,[15] the research aims to bridge materials science, digital fabrication, and architectural design, challenging conventional construction practices and proposing new environmentally friendly architectural solutions.

Tailored Fiber Placement

Tailored Fiber Placement

TFP is a digital textile manufacturing method that is based on stitching continuous fibers to a base material. The technique allows precise placement of the fibers, for example, along stress lines which can maximize structural performance.[16] The stitched TFP preforms are then infused with resin in vacuum-assisted process and shaped in the mold.

By replacing the epoxy resin, commonly used in this process, with chitosan gels, this study aims to achieve a fully biodegradable fiber-reinforced composite that combines environmental sustainability with high material performance. The research focused on material formulation, drying, and forming methods of flax – chitosan composites. Two case studies were conducted in collaboration with students participating in the BioMat design studio and seminar. These case studies explored innovative applications of flax-chitosan composites in architecture and design, demonstrating the material's potential in creating eco-friendly, functional structures and products.

Chitosan Facade Panels

Design

The design of the panel was strongly influenced by the fabrication method and impregnation procedure. Since the chitosan-impregnated TFP preform must be dried with access to air, it was decided on a tensile structure that would not need a mold and would be formed on the frame. The flow of air allowed the solvent to evaporate and reach the final rigid form of the panel.

The dimensions of the panel were defined by the working surface of the TFP machine and should be around 50 × 50 cm. The workflow started with defining the total dimensions of the panel of the hexagon with a radius of 25 cm. The panel features a modular design, allowing it to connect with other modules through its corners and extend the facade system in all directions. The modularity also enables scalable configurations, making the system adaptable for various applications.

The design of the panel was recomputed and resulted in a refined mesh. The double curved surface was form-found with Kangaroo in GH Rhino. Structural analysis performed with Karamba provided the pattern of principal stress lines and the boundary curves of the mesh. These were flattened onto the World XY Plane. To finalize the fabrication path for TFP, additional post-processing in CAD software was necessary to incorporate features such as connection loops (Figure 13.1b).

Fabrication

TFP pattern was stitched on a jute fabric with flax fiber 1000 tex. A 5% chitosan solution, prepared by dissolving chitosan (deacetylation value 75.0–85.0%, viscosity 5–20 mPa·s; supplier: TCI Deutschland GmbH) in 20% acetic acid, was used for the impregnation of TFP preforms (Figure 13.2a). Due to the low viscosity of the chitosan solution, the TFP preform was first submerged in the mixture. As the solvent partially evaporated, thus increasing the

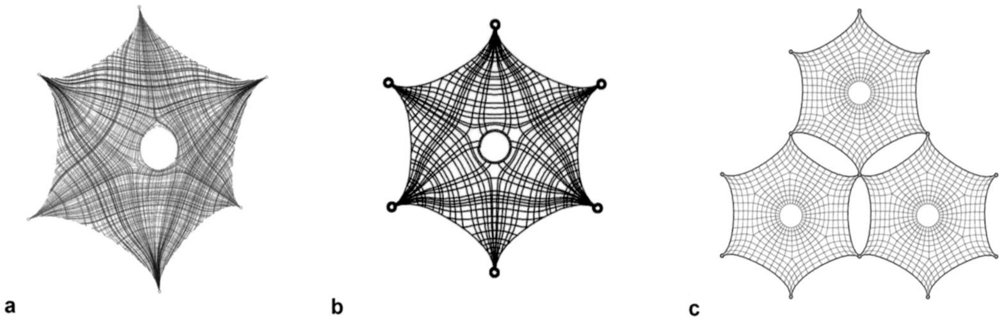

FIGURE 13.1 Chitosan facade panel: (a) principal stress lines (b) TFP pattern (c) modular design. Drawings created by design studio participants.

chitosan's concentration, the chitosan solution turned into a viscous gel matrix (Figure 13.2b). At this stage, the impregnated TFP-preform, could be stretched on the frame, where it stayed till the end of drying (Figure 13.2c). To speed up the drying process, the TFP preform was put in the oven for 12 h at 40°C. Using the same frame, a total of 4 panels were made and installed in an exemplary facade design (Figure 13.2d,e).

Chitoscopic Stool – Self Shaping of Chitosan Biocomposites

Design

The second TFP-project investigated self-shaping as an alternative shaping technique for chitosan-flax biocomposites. In this context, self-shaping refers to the controlled deformation that occurs during the evaporation of acetic acid. Initial investigations focused on smaller-scale studies, revealing a significant relationship between fiber spacing and curvature radius. It was also observed that by stitching both sides of the jute base fabric, double curvature can be achieved (Figure 13.3a). Based on the analysis of the measured data, a logarithmic behavior could be identified, showing the correlation between the fiber spacing and the curvature after drying (Figure 13.3b). Building upon these findings, a stool design was developed that uses the parallel fiber pattern to self-fold into a closed corrugated cross-section (Figures 13.4 & 13.5). Different cross-section profiles, created using various radius combinations, were analyzed with FEM structural analysis in Karamba. The final fiber spacings were determined by testing different radii and optimizing to minimize displacement and ranged from 6.3 mm to 9.0 mm, correlating to a radius of 20 mm to 30 mm (Figure 13.4).

Fabrication

A natural woven fabric was used as the base material for stitching, reinforced with a pattern made of continuous natural flax fibers with a TEX of 1300. TFP-Preforms were infused using a vacuum-assisted process with a 9% chitosan solution. The higher concentration of the solution provided a higher viscosity, which facilitated and accelerated the drying process. The preforms were then removed from the vacuum bag and were dried on an open rack to ensure

FIGURE 13.2 Chitosan facade panels fabrication: (a) chitosan gel (b) impregnated TFP preform (c) forming on the frame (d),(e) Chitosan Facade prototype. The collage includes images created by the author and design studio participants. Collage compiled by the author.

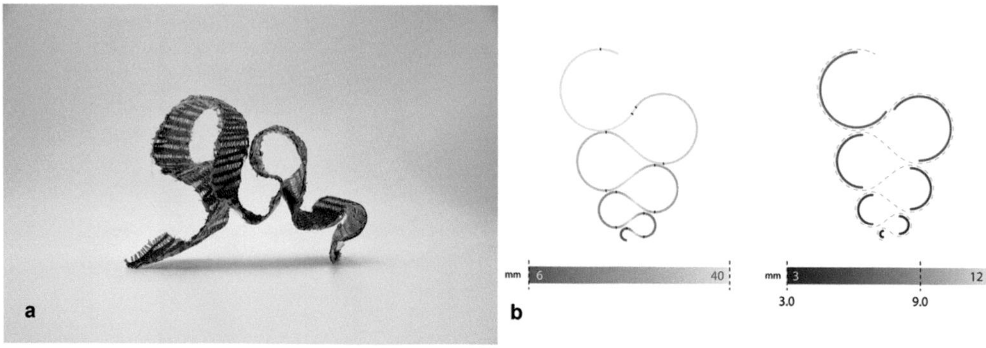

FIGURE 13.3 Self shaping studies: (a) corrugation test sample (b) Radius and correlating fiber spacing. Photograph and drawing created by MML seminar participants.

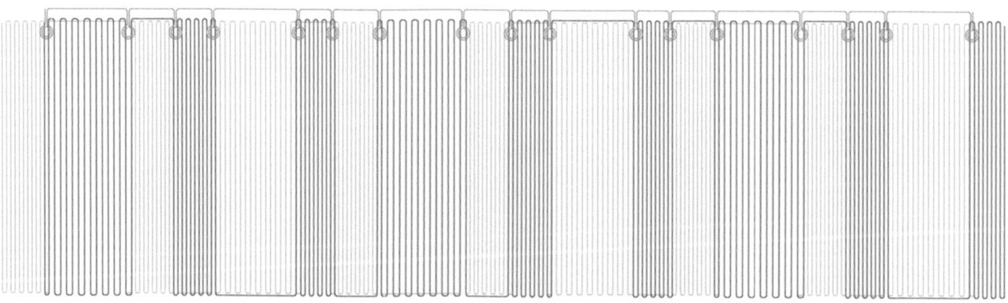

FIGURE 13.4 TFP pattern for the stool. Drawing created by MML seminar participants.

even drying on both sides. Once fully dried, the wooden seat was attached using a hemp rope threaded through pre-made holes in both the composite base and the seat.

The stool, measuring 40 cm in height and 35 cm in width, reached its final form after 8 days of drying, exhibiting slight deviations from the originally planned geometry. Weighing just 256 g, the chitosan fiber stool successfully supported the full weight of a 70 kg adult and a 1 kg wooden seat. The experimental stool serves as a proof of concept, demonstrating the potential for creating complex geometries and the practical application of self-shaping techniques of chitosan-flax biocomposites.

Results

The two TFP projects presented above provided important information regarding the fabrication of chitosan-flax composites. A comparison of two impregnation and forming methods reveals that vacuum infusion is more efficient than the soaking in a chitosan solution method. Vacuum infusion results in faster drying, reduced residual chitosan solution, and offers greater control over the amount of chitosan applied. Additionally, this method improves fiber impregnation. In terms of forming methods, the self-shaping approach requires further investigation, particularly concerning scalability and precision. In contrast, the frame-based forming method maintains fiber tension during drying, which is essential for load distribution, strength, stiffness, and

FIGURE 13.5 Chitoscopic stool: cross section profile and the final prototype. Photographs by MML seminar participants.

stability of the final product. This method is however limited in terms of geometry and is only suitable for specific applications.

Despite these limitations, the both case studies prove the chitosan-based fluids can effectively replace traditional resins while maintaining conventional composite manufacturing methods. As demonstrated in both projects, the integration of long flax fibers improved force distribution, leading to increased strength and stiffness of the composite, exhibiting load-bearing capabilities. While the mechanical properties require further investigation, the results underscore the potential of integrating lightweight biodegradable composite materials in architectural elements without compromising structural integrity.

3D Printing with Chitosan Paste

3D Printing

3D printing of biocomposites is a growing field that combines the advantages of additive manufacturing with sustainable and biodegradable materials. There are various 3D printing technologies used in the fabrication of biocomposites, each offering unique benefits depending on the application and required material properties.[17] One of the technologies particularly suitable for large-scale architectural applications is Liquid Deposition Modeling (LDM), also referred to as cold extrusion.[18] It is a printing technology, where the material in paste form is extruded through the nozzle at room temperature (e.g., using air pressure), and solidifies after deposition. This technique is commonly used to print materials such as clay, concrete, mycelium, and, more recently, composites.[19] This project explores the use of chitosan-based paste as a novel, sustainable material for LDM and its potential application in architectural contexts.

Chitosan Paste Development

The paste is prepared by mixing high-concentration chitosan gel with natural fibers and natural thickening agents (Figure 13.6). Chitosan acts as a matrix of the paste, which solidifies after the evaporation of acetic acid and binds all the ingredients. The natural fibers like flax or cotton,

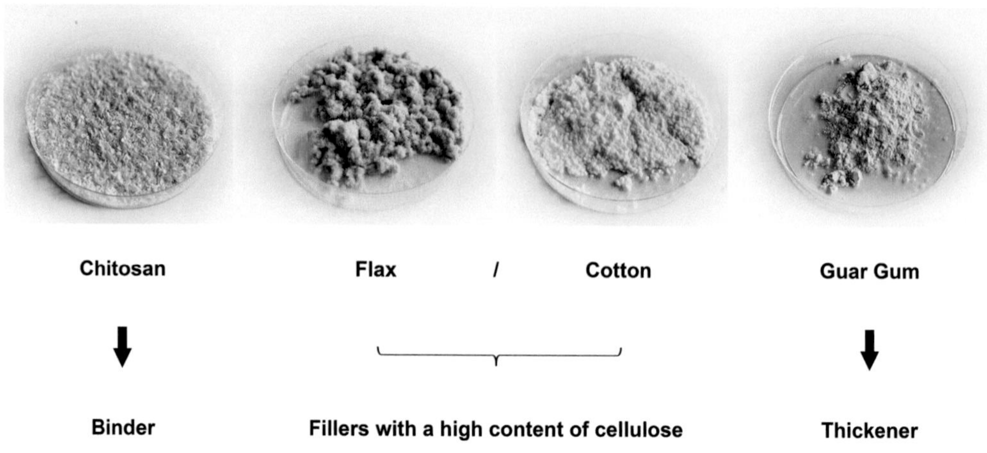

FIGURE 13.6 Chitosan paste ingredients. Photographs and collage by the author.

obtained from a local producer, were selected as the filler owing to its high cellulose content, which promotes robust binding interactions with chitosan. To enhance the paste's rheological properties, guar gum, a natural thickening agent derived from guar beans, was incorporated. All the materials were chosen for their natural origin, biodegradability and compatibility, aiming to optimize the printability and mechanical performance of the paste.

The formulation of the biocomposite chitosan paste was developed by testing various proportions of chitosan, fibers, and guar gum. Initial experiments helped identify the best proportions of 40% chitosan, 50% fibers, and 10% guar gum (utilizing 10% chitosan solution and 10% acetic acid as a solvent) (Figure 13.7). Printability and shrinkage were the primary factors considered during this optimization process. In addition to formulation, the mechanical and physical properties of the paste were thoroughly investigated. Compression tests revealed that the paste exhibits properties comparable to clay bricks, with a maximum compressive strength of 21.7 MPa and a middle value of 15.1 MPa, suggesting its potential application in specific architectural contexts.[20] Given the hydrophilic nature of chitosan, additionally water resistance tests were conducted. The results demonstrated that when coated with natural substances like beeswax, the biocomposite becomes resistant to water absorption.

Architectural Application

Having a comprehensive understanding of the materials' properties, geometrical studies and architectural scenarios were developed and prototypes were fabricated. Given the material's favorable compressive strength, several concepts for 3D printed off-site modular and on-site monolithic walls were created.

Modular Walls

Modular systems benefit from off-site fabrication, which is especially advantageous for chitosan printing where the controlled environment is necessary for both printing and drying processes. The integration of 3D printing enables the customization and design flexibility of individual bricks or modules. Compared to traditional bricks, 3D-printed chitosan bricks can offer several

FIGURE 13.7 Chitosan paste preparation: (1) chitosan solution, (2) mixing with flax fibers, (3) ready paste, (4) 3D printing process. Photographs by the A. Brückner. Collage compiled by the author.

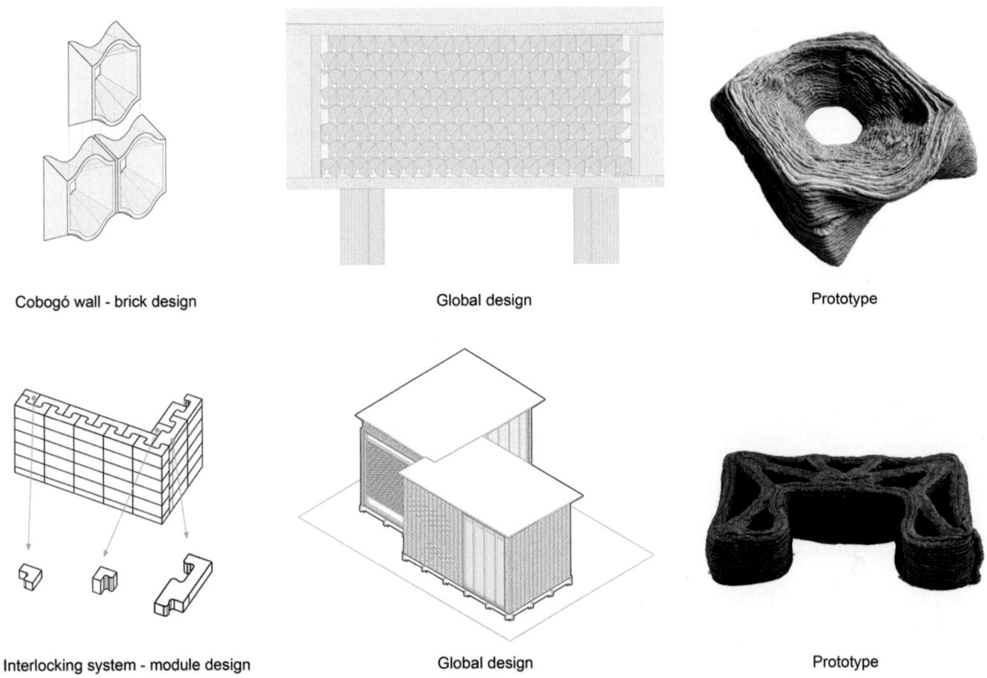

FIGURE 13.8 Modular wall designs and printed prototypes: Cobogó Wall (top row), Interlocking Modular Wall System (bottom row). The collage includes images created by the author and design studio participants. Collage compiled by the author.

advantages, particularly in terms of sustainability. Made from renewable and biodegradable materials, these bricks offer a lightweight, environmentally friendly alternative to traditional clay bricks, requiring less energy for production as they eliminate the need for kiln firing.[21]

The first design uses the advantages of 3D printing technology to create a set of unique, custom-designed chitosan bricks. Inspired by the Pedregulho Housing Complex, the porous facade design uses cobogó elements to enhance shading, ventilation, and passive climate control through the Venturi effect. The bricks were designed using Grasshopper3D, a parametric design software, and were optimized for their specific position within the wall using the Ladybug Plugin, providing optimal solar protection. The combination of parametric design and 3D printing allows for quick adaptation to different parameters, applications, and sites (Figure 13.8).

The second design focuses on creating a modular system for temporary structures. By incorporating interlocking modules, the system simplifies assembly and disassembly, which significantly reduces construction time and complexity. Material formulations and printing paths are optimized to enhance printability and structural integrity. The design is highly adaptable, allowing for future modifications and extensions, making it a cost-effective and environmentally friendly solution for temporary or evolving structures.

Both designs exemplify the versatility and sustainable advantages of 3D-printed modular chitosan wall systems. The lightweight bricks can be assembled using either chitosan paste or removable connectors, facilitating adaptability, reusability, and efficient construction.

Monolithic Walls

Expanding the versatility of the material's application, additional scenarios for monolithic 3D-printed on-site partition and structural walls were considered. On-site printing simplifies construction by eliminating shipping, allowing faster fabrication. However, printing larger volumes presents printing and drying challenges, which must be addressed in the design.

The first wall is a lattice structure designed to optimize efficiency and sustainability (Figure 13.9). This lattice design distributes loads evenly, ensuring structural integrity while minimizing material usage and reducing the overall weight. Additionally, the lattice structure provides a larger surface area, which accelerates and equalizes the drying process, reducing deformation risks and ensuring good quality of the final product.

The second design features a straight wall integrated with an infill pattern optimized for efficiency and functionality. This pattern facilitates efficient drying and ensures smooth, uninterrupted printing through a continuous path. Additionally, the design accommodates the integration of chitosan-based insulation material, which is currently under investigation within the ChitinFluid research project.

Prototypes Fabrication

The printing of the prototypes was conducted using the DELTA WASP 2040 Clay printer, and following printing parameters: 3 mm nozzle, printing speed 60–150 mm/s, 100% flow and pressure from 3 to 6 bars (provided by the additional compressor). All the prototypes were printed using custom G-codes generated in Grasshopper, enabling continuous, smooth, and clean printing. Due to the shrinkage that occurs during drying, the models had to be scaled in z-axis by 1/3 which corresponds to the volume of acetic acid in the paste, which evaporates.

All the prototypes were dried at room temperature and reached the expected height, however, some of them have deformed weather during printing or drying process. The observed printing defects can be categorized into two types. The first type results from the gravitational stress applied by the upper layers, causing the lower part to expand slowly. This issue can be prevented by speeding up the drying process, for example, by using fans. The second category of defects involves vertical deformation, which can be prevented by having the infill pattern that supports the outer walls. Overall, a continuous printing path combined with geometry that enables air circulation and ensures stability during printing, can significantly improve both printing quality and the precision of the final product (Figure 13.10).

FIGURE 13.9 Scenarios for printing on-site: lattice wall (top row), wall with optimized infill pattern (bottom row). The collage includes images created by the author and seminar participants. Collage compiled by the author.

FIGURE 13.10 3D printed prototypes. Photograph by the author.

Results

In summary, the biocomposite paste based on chitosan and natural fibers requires further improvement in mechanical strength and shrinkage. Optimizing drying methods for larger components should be a key area for future research. The hydrophilic nature of chitosan and the necessity of protective coating currently limit its broader application in architecture. Although the paste and process require further optimization, the findings show that chitin-based paste has a potential in eco-friendly construction and could be a sustainable alternative to conventional materials. 3D printed chitosan walls offer significant advantages in sustainability, reduced weight, and high strength-to-weight ratio, which can reduce structural loads and simplify construction. These properties make them suitable for applications such as temporary structures, interior partitions, low-rise buildings, and other architectural scenarios with smaller load requirements. Modular design and prefabrication, which allow for better-controlled manufacturing environments, seem to be the most feasible approach for large-scale application of chitosan paste. Further research should focus on optimizing the paste and scaling up the process, as well as improving water resistance by replacing chitosan with chitin to enhance durability and the possibility of application in diverse architectural uses.

Summary

The work presented in this chapter demonstrates the significant potential of chitosan-based composites to create advanced sustainable architecture and design. The integration of TFP and 3D printing techniques with biodegradable materials provides an innovative approach to creating high-performance, eco-friendly architectural elements. Despite the current limitations, the findings suggest promising pathways for the future development of chitosan- and chitin-based materials and their applicability in sustainable construction.

Acknowledgements

The work presented in this paper is part of the results of the research project "Chitin fluid: Chitin as a Resource for Multifunctional Materials via Water-Based Complex Fluids," funded by the Carl Zeiss Stiftung. The Chitosan Facade Panels were developed in a Design Studio 2021–2022, offered by BioMat by students S. D. Viersbach Gordillo, J. Schwarzenbacher, and M. Thi Nguyen under the tutoring of J. Petrš and P. Grabowska, and the supervision of H. Dahy. The Chitoscopic Stool was developed in the Material Matter Lab (MML) seminar 2022–2023, by students M. Fouillat and L. Neumann under the tutoring of P. Grabowska and the supervision of H. Dahy. The Cobogó Wall was developed in a Design Studio in 2023, offered by BioMat by students L. Rédua Bispo, A. Kurumaz, and Ž. Potnik, under the tutoring of P. Grabowska and the supervision of H. Dahy. The Interlocking Modular Wall System was developed as part of master's thesis by Y. Zho, under the tutoring of P. Grabowska and the supervision of H. Dahy. The Lattice Wall was developed in the Material Matter Lab (MML) seminar 2022–2023, by students L. Tiedemann, A. Huynh, and F. Ruff, under the tutoring of P. Grabowska and the supervision of H. Dahy.

During the preparation of this work, the authors used ChatGPT as a tool to improve readability, revise grammar, and enhance language. After utilizing this tool, the authors reviewed and edited the content as needed and take full responsibility for the content of the publication.

Notes

1 Yale Center for Ecosystems + Architecture United Nations Environment Programme, "Building Materials and the Climate: Constructing a New Future," September 2023, https://wedocs.unep.org/20.500.11822/43293.
2 Jorge Duro-Royo et al., "Designing a Tree: Fabrication Informed Digital Design and Fabrication of Hierarchical Structures," in *Proceedings of IASS Annual Symposia*, vol. 2018 (International Association for Shell and Spatial Structures (IASS), 2018), 1–7; Naresh D. Sanandiya et al., "Circular Manufacturing of Chitinous Bio-Composites via Bioconversion of Urban Refuse," *Scientific Reports* 10, no. 1 (March 13, 2020): 4632, https://doi.org/10.1038/s41598-020-61664-1; Niaz Mahmud, Joinul Islam, and Reza Tahergorabi, "Marine Biopolymers: Applications in Food Packaging," *Processes* 9, no. 12 (December 13, 2021): 2245, https://doi.org/10.3390/pr9122245; Manuel Kretzer and Sina Mostafavi, "3. Bioplastic Robotic Materialization. Design to Robotic Production of Biodegradable Lamps," *ICS MATERIALS*, 2021; Resego Phiri et al., "Development of Sustainable Biopolymer-Based Composites for Lightweight Applications from Agricultural Waste Biomass: A Review," *Advanced Industrial and Engineering Polymer Research* 6, no. 4 (October 2023): 436–50, https://doi.org/10.1016/j.aiepr.2023.04.004; Materiability Research Group, "Microbial Leather," Web article, Project Page, n.d., https://materiability.com/portfolio/microbial-leather/.
3 Phiri et al., "Development of Sustainable Biopolymer-Based Composites for Lightweight Applications from Agricultural Waste Biomass"; LALIT Ranakoti, MAYANK Pokhriyal, and ANKUR Kumar, "Natural Fibers and Biopolymers Characterization: A Future Potential Composite Material," *Journal of Mechanical Engineering* 68, no. 1 (2018): 33–50.
4 Bryan Harris, "Engineering Composite Materials," 1999.
5 Hanaa Dahy, "Materials as Design Tool: Digital Fabrication Strategies for Sustainable Architecture," *Technology|Architecture + Design* 7, no. 2 (July 3, 2023): 153–58, https://doi.org/10.1080/24751448.2023.2245707; Gabriel Rihaczek et al., "Curved Foldable Tailored Fiber Reinforcements for Moldless Customized Bio-Composite Structures. Proof of Concept: Biomimetic NFRP Stools," *Polymers* 12, no. 9 (September 2, 2020): 2000, https://doi.org/10.3390/polym12092000; Piotr Baszyński and Hanaa Dahy, "Free-Forming of Customised NFRP Profiles for Architecture Using Simplified Adaptive and Stay-In-Place Moulds," *Designs* 8, no. 6 (December 3, 2024): 129, https://doi.org/10.3390/designs8060129; Hanaa Dahy, "Natural Fibre-Reinforced Polymer Composites (NFRP) Fabricated from Lignocellulosic Fibres for Future Sustainable Architectural Applications, Case Studies: Segmented-Shell Construction, Acoustic Panels, and Furniture," *Sensors* 19, no. 3 (February 12, 2019): 738, https://doi.org/10.3390/s19030738.
6 S. Vigneshwaran et al., "Recent Advancement in the Natural Fiber Polymer Composites: A Comprehensive Review," *Journal of Cleaner Production* 277 (December 2020): 124109, https://doi.org/10.1016/j.jclepro.2020.124109.
7 Marguerite Rinaudo, "Chitin and Chitosan: Properties and Applications," *Progress in Polymer Science* 31, no. 7 (July 2006): 603–32, https://doi.org/10.1016/j.progpolymsci.2006.06.001.
8 Rinaudo; Vanessa P. Santos et al., "Seafood Waste as Attractive Source of Chitin and Chitosan Production and Their Applications," *International Journal of Molecular Sciences* 21, no. 12 (June 16, 2020): 4290, https://doi.org/10.3390/ijms21124290.
9 Anran Wei, Jimin Fu, and Fenglin Guo, "Mechanical Properties of Chitin Polymorphs: A Computational Study," *Journal of Materials Science* 56, no. 20 (July 2021): 12048–58, https://doi.org/10.1007/s10853-021-06086-8.
10 Rinaudo, "Chitin and Chitosan."
11 Rinaudo; Ioannis A. Sogias, Vitaliy V. Khutoryanskiy, and Adrian C. Williams, "Exploring the Factors Affecting the Solubility of Chitosan in Water," *Macromolecular Chemistry and Physics* 211, no. 4 (February 15, 2010): 426–33, https://doi.org/10.1002/macp.200900385.
12 Antje Ota et al., "Chitin/Cellulose Blend Fibers Prepared by Wet and DRY-WET Spinning," *Polymers for Advanced Technologies* 32, no. 1 (January 2021): 335–42, https://doi.org/10.1002/pat.5089.
13 Sanandiya et al., "Circular Manufacturing of Chitinous Bio-Composites via Bioconversion of Urban Refuse"; Naresh D. Sanandiya et al., "Large-Scale Additive Manufacturing with Bioinspired Cellulosic Materials," *Scientific Reports* 8, no. 1 (June 5, 2018): 8642, https://doi.org/10.1038/s41598-018-26985-2.

14 Furqan Ahmad, Heung Soap Choi, and Myung Kyun Park, "A Review: Natural Fiber Composites Selection in View of Mechanical, Light Weight, and Economic Properties," *Macromolecular Materials and Engineering* 300, no. 1 (January 2015): 10–24, https://doi.org/10.1002/mame.201400089.

15 Hanaa Dahy, " 'Materials as a Design Tool' Design Philosophy Applied in Three Innovative Research Pavilions Out of Sustainable Building Materials with Controlled End-Of-Life Scenarios," *Buildings* 9, no. 3 (March 13, 2019): 64, https://doi.org/10.3390/buildings9030064.

16 P.J. Crothers et al., "Tailored Fibre Placement to Minimise Stress Concentrations," *Composites Part A: Applied Science and Manufacturing* 28, no. 7 (January 1997): 619–25, https://doi.org/10.1016/S1359-835X(97)00022-5.

17 Antoine Le Duigou et al., "A Review of 3D and 4D Printing of Natural Fibre Biocomposites," *Materials & Design* 194 (September 2020): 108911, https://doi.org/10.1016/j.matdes.2020.108911; Jun Liu et al., "Current Advances and Future Perspectives of 3D Printing Natural-Derived Biopolymers," *Carbohydrate Polymers* 207 (March 2019): 297–316, https://doi.org/10.1016/j.carbpol.2018.11.077; Vasi Uddin Siddiqui et al., "Sustainable Biocomposites," in *Additive Manufacturing for Biocomposites and Synthetic Composites*, by M. T. Mastura, S. M. Sapuan, and R. A. Ilyas, 1st ed. (Boca Raton: CRC Press, 2023), 1–19, https://doi.org/10.1201/9781003362128-1.

18 Elise Elsacker, Eveline Peeters, and Lars De Laet, "Large-Scale Robotic Extrusion-Based Additive Manufacturing with Living Mycelium Materials," *Sustainable Futures* 4 (2022): 100085, https://doi.org/10.1016/j.sftr.2022.100085.

19 Arnaldo D. Valino et al., "Advances in 3D Printing of Thermoplastic Polymer Composites and Nanocomposites," *Progress in Polymer Science* 98 (November 2019): 101162, https://doi.org/10.1016/j.progpolymsci.2019.101162; Elsacker, Peeters, and De Laet, "Large-Scale Robotic Extrusion-Based Additive Manufacturing with Living Mycelium Materials"; Andrei Goidea, Dimitrios Floudas, and Daniel Andréen. "Pulp Faction: 3d Printed Material Assemblies through Microbial Biotransformation." In *Fabricate 2020: Making Resilient Architecture*, edited by J. Burry, J. Sabin, B. Sheil, and M. Skavara, 42–49. London: UCL Press, 2020.

20 Anant L. Murmu and A. Patel, "Towards Sustainable Bricks Production: An Overview," *Construction and Building Materials* 165 (March 2018): 112–25, https://doi.org/10.1016/j.conbuildmat.2018.01.038.

21 Murmu and Patel, "Towards Sustainable Bricks Production."

References

Ahmad, Furqan, Heung Soap Choi, and Myung Kyun Park. "A Review: Natural Fiber Composites Selection in View of Mechanical, Light Weight, and Economic Properties." *Macromolecular Materials and Engineering* 300, no. 1 (January 2015): 10–24. https://doi.org/10.1002/mame.201400089.

Baszyński, Piotr, and Hanaa Dahy. "Free-Forming of Customised NFRP Profiles for Architecture Using Simplified Adaptive and Stay-In-Place Moulds." *Designs* 8, no. 6 (December 3, 2024): 129. https://doi.org/10.3390/designs8060129.

Crothers, P. J., K. Drechsler, D. Feltin, I. Herszberg, and T. Kruckenberg. "Tailored Fibre Placement to Minimise Stress Concentrations." *Composites Part A: Applied Science and Manufacturing* 28, no. 7 (January 1997): 619–25. https://doi.org/10.1016/S1359-835X(97)00022-5.

Dahy, Hanaa. "'Materials as a Design Tool' Design Philosophy Applied in Three Innovative Research Pavilions Out of Sustainable Building Materials with Controlled End-Of-Life Scenarios." *Buildings* 9, no. 3 (March 13, 2019): 64. https://doi.org/10.3390/buildings9030064.

Dahy, Hanaa. "Materials as Design Tool: Digital Fabrication Strategies for Sustainable Architecture." *Technology|Architecture + Design* 7, no. 2 (July 3, 2023): 153–58. https://doi.org/10.1080/24751448.2023.2245707.

Dahy, Hanaa. "Natural Fibre-Reinforced Polymer Composites (NFRP) Fabricated from Lignocellulosic Fibres for Future Sustainable Architectural Applications, Case Studies: Segmented-Shell Construction, Acoustic Panels, and Furniture." *Sensors* 19, no. 3 (February 12, 2019): 738. https://doi.org/10.3390/s19030738.

Duro-Royo, Jorge, Josh Van Zak, Andrea Ling, Yen-Ju Tai, Nicolas Hogan, Barrak Darweesh, and Neri Oxman. "Designing a Tree: Fabrication Informed Digital Design and Fabrication of Hierarchical Structures." In *Proceedings of IASS Annual Symposia, 2018*: 1–7. International Association for Shell and Spatial Structures (IASS), 2018.

Elsacker, Elise, Eveline Peeters, and Lars De Laet. "Large-Scale Robotic Extrusion-Based Additive Manufacturing with Living Mycelium Materials." *Sustainable Futures* 4 (2022): 100085. https://doi.org/10.1016/j.sftr.2022.100085.

Goidea, Andrei, Dimitrios Floudas, and Daniel Andréen. "Pulp Faction: 3d Printed Material Assemblies through Microbial Biotransformation." In *Fabricate 2020: Making Resilient Architecture*, edited by J. Burry, J. Sabin, B. Sheil, and M. Skavara, 42–49. London: UCL Press, 2020.

Harris, Bryan. "Engineering Composite Materials." London: Institute of Materials, 1999.

Kretzer, Manuel, and Sina Mostafavi. "3. Bioplastic Robotic Materialization. Design to Robotic Production of Biodegradable Lamps." *ICS MATERIALS*, 2021: 34–46.

Le Duigou, Antoine, David Correa, Masahito Ueda, Ryosuke Matsuzaki, and Mickael Castro. "A Review of 3D and 4D Printing of Natural Fibre Biocomposites." *Materials & Design* 194 (September 2020): 108911. https://doi.org/10.1016/j.matdes.2020.108911.

Liu, Jun, Lushan Sun, Wenyang Xu, Qianqian Wang, Sujie Yu, and Jianzhong Sun. "Current Advances and Future Perspectives of 3D Printing Natural-Derived Biopolymers." *Carbohydrate Polymers* 207 (March 2019): 297–316. https://doi.org/10.1016/j.carbpol.2018.11.077.

Mahmud, Niaz, Joinul Islam, and Reza Tahergorabi. "Marine Biopolymers: Applications in Food Packaging." *Processes* 9, no. 12 (December 13, 2021): 2245. https://doi.org/10.3390/pr9122245.

Materiability Research Group. "Microbial Leather." Web article, Project Page, n.d. https://materiability.com/portfolio/microbial-leather/.

Murmu, Anant L., and A. Patel. "Towards Sustainable Bricks Production: An Overview." *Construction and Building Materials* 165 (March 2018): 112–25. https://doi.org/10.1016/j.conbuildmat.2018.01.038.

Ota, Antje, Ronald Beyer, Ulrich Hageroth, Alexandra Müller, Patricija Tomasic, Frank Hermanutz, and Michael R. Buchmeiser. "Chitin/Cellulose Blend Fibers Prepared by Wet and DRY-WET Spinning." *Polymers for Advanced Technologies* 32, no. 1 (January 2021): 335–42. https://doi.org/10.1002/pat.5089.

Phiri, Resego, Sanjay Mavinkere Rangappa, Suchart Siengchin, Oluseyi Philip Oladijo, and Hom Nath Dhakal. "Development of Sustainable Biopolymer-Based Composites for Lightweight Applications from Agricultural Waste Biomass: A Review." *Advanced Industrial and Engineering Polymer Research* 6, no. 4 (October 2023): 436–50. https://doi.org/10.1016/j.aiepr.2023.04.004.

Ranakoti, Lalit, Mayank Pokhriyal, and Ankur Kumar. "Natural Fibers and Biopolymers Characterization: A Future Potential Composite Material." *Journal of Mechanical Engineering* 68, no. 1 (2018): 33–50.

Rihaczek, Gabriel, Maximilian Klammer, Okan Başnak, Jan Petrš, Benjamin Grisin, Hanaa Dahy, Stefan Carosella, and Peter Middendorf. "Curved Foldable Tailored Fiber Reinforcements for Moldless Customized Bio-Composite Structures. Proof of Concept: Biomimetic NFRP Stools." *Polymers* 12, no. 9 (September 2, 2020): 2000. https://doi.org/10.3390/polym12092000.

Rinaudo, Marguerite. "Chitin and Chitosan: Properties and Applications." *Progress in Polymer Science* 31, no. 7 (July 2006): 603–32. https://doi.org/10.1016/j.progpolymsci.2006.06.001.

Sanandiya, Naresh D., Christoph Ottenheim, Jun Wei Phua, Augusta Caligiani, Stylianos Dritsas, and Javier G. Fernandez. "Circular Manufacturing of Chitinous Bio-Composites via Bioconversion of Urban Refuse." *Scientific Reports* 10, no. 1 (March 13, 2020): 4632. https://doi.org/10.1038/s41598-020-61664-1.

Sanandiya, Naresh D., Yadunund Vijay, Marina Dimopoulou, Stylianos Dritsas, and Javier G. Fernandez. "Large-Scale Additive Manufacturing with Bioinspired Cellulosic Materials." *Scientific Reports* 8, no. 1 (June 5, 2018): 8642. https://doi.org/10.1038/s41598-018-26985-2.

Santos, Vanessa P., Nathália S. S. Marques, Patrícia C. S. V. Maia, Marcos Antonio Barbosa De Lima, Luciana De Oliveira Franco, and Galba Maria De Campos-Takaki. "Seafood Waste as Attractive Source of Chitin and Chitosan Production and Their Applications." *International Journal of Molecular Sciences* 21, no. 12 (June 16, 2020): 4290. https://doi.org/10.3390/ijms21124290.

Siddiqui, Vasi Uddin, S.M. Sapuan, Sulaiman Ahmad, Luqman Hakim Noorazlan, Neassan Muthamil Selvan, Shudharson Ilengo, Pavithren Mailvaganan, and Vairravell Khumar Ravindran. "Sustainable

Biocomposites." In *Additive Manufacturing for Biocomposites and Synthetic Composites*, edited by M. T. Mastura, S. M. Sapuan, and R. A. Ilyas, 1–19, 1st ed. Boca Raton: CRC Press, 2023. https://doi.org/10.1201/9781003362128-1.

Sogias, Ioannis A., Vitaliy V. Khutoryanskiy, and Adrian C. Williams. "Exploring the Factors Affecting the Solubility of Chitosan in Water." *Macromolecular Chemistry and Physics* 211, no. 4 (February 15, 2010): 426–33. https://doi.org/10.1002/macp.200900385.

United Nations Environment Programme, Yale Center for Ecosystems + Architecture. "Building Materials and the Climate: Constructing a New Future." September 2023. https://wedocs.unep.org/20.500.11822/43293.

Valino, Arnaldo D., John Ryan C. Dizon, Alejandro H. Espera, Qiyi Chen, Jamie Messman, and Rigoberto C. Advincula. "Advances in 3D Printing of Thermoplastic Polymer Composites and Nanocomposites." *Progress in Polymer Science* 98 (November 2019): 101162. https://doi.org/10.1016/j.progpolymsci.2019.101162.

Vigneshwaran, S., R. Sundarakannan, K.M. John, R. Deepak Joel Johnson, K. Arun Prasath, S. Ajith, V. Arumugaprabu, and M. Uthayakumar. "Recent Advancement in the Natural Fiber Polymer Composites: A Comprehensive Review." *Journal of Cleaner Production* 277 (December 2020): 124109. https://doi.org/10.1016/j.jclepro.2020.124109.

Wei, Anran, Jimin Fu, and Fenglin Guo. "Mechanical Properties of Chitin Polymorphs: A Computational Study." *Journal of Materials Science* 56, no. 20 (July 2021): 12048–58. https://doi.org/10.1007/s10853-021-06086-8.

14

TOWARD NATURAL FIBER ARCHITECTURE

From Synthetic to Bio-Based Composite Building Systems

Christoph Zechmeister, Monika Göbel, Jan Knippers, and Achim Menges

Composites in Technology and Architecture

While textile and membrane structures have shaped human shelter for millennia, fiber-reinforced polymer composites – created by embedding fibrous elements such as glass or carbon into a polymer matrix – were originally developed for industrial applications in aviation and maritime contexts. Valued for their high strength-to-weight ratio, corrosion resistance, and durability, they served primarily as substitutes for traditional materials like metal, but are lighter, cheaper, and often faster to produce.[1]

In architecture, experimental projects of the 1950s and 1960s – such as the Monsanto House or the Futuro House – imagined a future of serially produced, site-independent dwellings.[2] Yet their promise was tempered by technical limitations and a lack of contextual sensitivity. In the following decades, fiber composites gained a foothold in architecture, particularly in non-load-bearing applications such as freeform cladding or bespoke facade panels – elements that benefited from the material's pliability but remained dependent on labor- and resource-intensive formwork.[3]

This reliance on molds – whether positive, negative, or mandrel-based – has profoundly shaped the architectural use of composites. Fabrication remains costly, wasteful, and largely limited to repetitive geometries or high-budget projects. Such constraints are increasingly incompatible with the urgent demands of sustainable construction. The potential of fiber composites in architecture is therefore not only tied to what they are made of, but also to how they are produced and how their inherent material behavior can be leveraged.[4]

Here, nature offers a compelling model. In most load-bearing natural composites, performance emerges from variation in fiber orientation, density, and arrangement, locally adapted to specific structural or functional demands.[5] Unlike conventional construction, nature exploits anisotropy and geometrical complexity rather than avoiding it. This dynamic use of material intelligence offers a rich reservoir of structural, morphological, and procedural principles that can inform a new architectural approach.

DOI: 10.4324/9781003613589-17

Plastic as a Continuous Process of Formation

Building on the concept of dynamic material behavior in fiber composites, coreless filament winding (CFW) breaks free from the traditional reliance on molds or mandrels. Instead, fibers are robotically deposited onto a series of anchor points arranged on spatial frames (Figure 14.1), forming complex, multi-layered structures through the dynamic, reciprocal interaction of the fibers themselves.[6]

This method treats plastic not as a static material but as an active process – a continuous shaping and reshaping unfolding in space over time. CFW's inherently non-linear and adaptive nature strongly depends on digital simulation and robotic fabrication, enabling the material to "form itself" in response to structural and geometric demands, guided by the robotic agent. Unlike conventional filament winding, which imposes a pre-determined shape on the material through its mold, coreless filament winding encourages morphological variation. The added control achieved by the design of different fiber patterns combined into a fiber layup provides the designer with a high degree of control over the aesthetics and performance of the component in local areas, allowing for specific customization, creating unique, load-adapted architectural components that optimize material expenditure and minimize production waste.[7]

The development of architectural applications for coreless winding unfolds across three deeply interconnected domains: robotic fabrication processes, computational design methods for pattern generation, and the integration of these into digital building systems. Here, the digital

FIGURE 14.1 Coreless filament winding process.

FIGURE 14.2 Full-scale architectural fiber composite demonstrators.

becomes not only a tool but a medium for integration, enabling the coordination of material behavior, fabrication constraints, and architectural intent.[8]

Over the past decade, this approach has been explored through a series of full-scale demonstrators developed by ICD and ITKE at the University of Stuttgart, each engaging a specific set of challenges posed by CFW (Figure 14.2). Early incremental investigations laid the groundwork for more complex, large-scale applications, demonstrating how a process-oriented understanding of composite formation might yield a new class of building systems – lightweight, materially responsive, and catering to a context increasingly defined by material scarcity.

Radical Material Savings, Ultra-High Performance

Two exemplary projects that illustrate the transformative potential of high-performance fibrous building elements are the BUGA Fibre Pavilion, a long-span dome structure, and Maison Fibre, a conceptual model for multi-story fiber structures. Both share a material system rooted in the precise combination of carbon and glass fibers. Carbon fiber, renowned for its exceptionally high mass-specific stiffness, is commonly used in high-performance applications such as aerospace, automotive, and other advanced engineering fields, while glass fiber offers a balance of high strength, cost-effectiveness, and greater environmental tolerance.

The BUGA Fibre Pavilion,[9] constructed for the Bundesgartenschau in Heilbronn, spans over 23 meters with a structure composed entirely of glass and carbon fibers – more than 150,000 meters in total (Figure 14.3). Sixty bespoke structural components were robotically fabricated without molds, using a reconfigurable boundary frame to define anchor points for the winding

FIGURE 14.3 The BUGA Fibre Pavilion, Bundesgartenschau Heilbronn 2019.

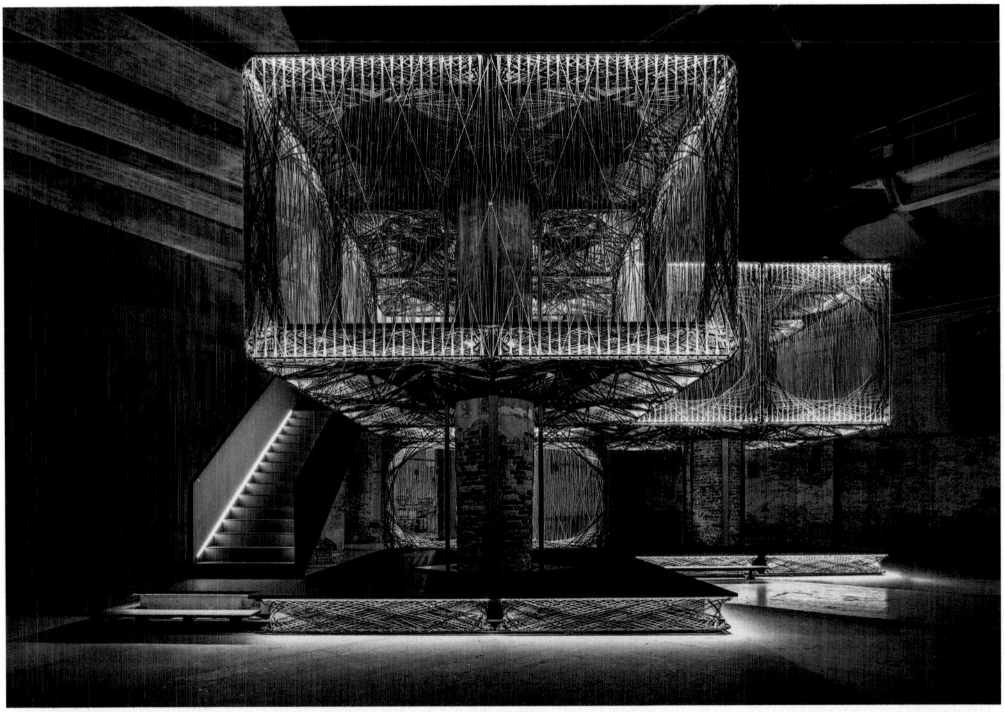

FIGURE 14.4 Maison Fibre, Venice Architecture Biennale 2021.

process. Glass fibers act as a translucent body lattice, while carbon fibers are selectively added in regions of high structural demand. This locally differentiated use of materials mirrors the efficiency of natural fiber systems, allowing strength and stiffness to be precisely tuned through fiber orientation and density. Despite its large scale, the pavilion's primary structure weighs only 7.6 kg per square meter – five times lighter than a comparable steel system – while the components withstand loads up to 250 kN. The interplay of transparent and opaque fibers reveals the logics of formation and performance, embedding structural articulation into architectural expression.

Maison Fibre,[10] exhibited at the Venice Architecture Biennale in 2021, extends CFW into a multi-story context for the first time, while maintaining the radical material savings enabled by a high-tech material system (Figure 14.4). The modular slab and wall system uses less than 2% of its volume in actual material. The upper floor structure, designed for a live load of 4 kN/m^2, weighs under 10 kg/m^2, and the side walls even less. This radical lightness is not only a structural achievement but an architectural one: the fibrous elements challenge the tectonic language of conventional building systems, offering a spatial expression of future, highly dematerialized structures, where each building element can be locally made from just a few kilos of construction material. Compact and adaptable, the installation as a conceptual model hints at future applications in on-site or modular construction, redefining how buildings might be expanded, have their function changed, or be renewed throughout their life cycle.

Both projects demonstrate a technological apex of synthetic composites in architecture, providing maximum performance and flexibility with a minimum of material expenditure through rigorous customization and tailoring to structural and functional requirements. Despite their unparalleled performance, high material efficiency, and minimal production waste, carbon fiber composites exhibit environmental and practical limitations. Their dependence on petrochemical resources results in high embodied energy and carbon footprints.[11] Additionally, epoxy resins are subject to degradation through UV exposure, which can lead not only to visual deterioration but also compromised structural performance. These issues have been observed in elements of the BUGA Pavilion and other outdoor demonstrators, where aging effects have become increasingly visible. Acknowledging these limitations is critical in evaluating the long-term viability of synthetic fiber systems.

From Synthetic to Provenance – Reframing Architectural Fiber Composites

Responding to the requirement to decarbonize construction, plant fibers like flax, hemp, or jute are emerging as viable alternatives to synthetic products such as carbon or glass fiber. Among these, flax fibers stand out due to their short cultivation cycle and local availability in Europe. Flax fibers, historically used in textile production, were central to Europe's linen industry before cotton became the dominant textile fiber.[12] They offer advantageous mechanical properties,[13] particularly high stiffness and strength, which have led to their increasing use in structural composites. Derived from annually renewable crops and cultivated using solar energy, flax fibers rely far less on fossil fuels throughout their growth and initial processing than conventional synthetic fibers. As a result, their life cycle exhibits a significantly lower dependency on non-renewable energy and materials, making them a promising material for high-performance, climate-conscious building systems. However, this potential is contingent on the broader ecological and industrial context. Cultivation is only truly sustainable in climatically suitable regions; otherwise, fiber extraction may lead to high water consumption or local water pollution. While fiber bundles used in architectural composites do not require spinning into yarns, this step dominates global natural fiber processing and is presently concentrated in China. If natural fibers are to be widely adopted in the building sector in Europe, a robust, localized value chain for cultivation, processing, and manufacturing will need to be reestablished.

Variability in mechanical properties, influenced by growing conditions, processing, and moisture sensitivity, presents additional challenges that must be addressed in both design and fabrication. Methods previously developed for synthetic composites must be expanded to embrace material uncertainty and integrate early-stage material characterization into the design process (Figure 14.5).

Performance is no longer the product of architectural intent, structural requirements, and manufacturing influence, but is significantly determined by material heterogeneity. This requires cyclical, feedback-driven workflows that treat material characterization, process calibration, and structural evaluation as interconnected and iterative. Flax fiber rovings are technically assembled products made from shorter-grown strands, showing irregular and non-standardized behavior that requires significant adaptations in robotic fabrication, particularly due to challenges in resin impregnation and reduced dry tensile strength. This requires the modification of fabrication infrastructure for a friction-reduced impregnation process, optimized tension control, and adjusted robot code for more precise fiber-tool alignment to avoid the disintegration of the delicate flax fiber bundle.[14]

As a first application of plant-based architectural composite structures, the livMatS Pavilion (Figure 14.6), located in the Botanical Garden of Freiburg, demonstrates this approach.[15] Conceived as a lightweight outdoor event structure, it consists of 15 load-bearing flax fiber

FIGURE 14.5 Material characterization process.

components radially arranged around a central keystone element. Each component, 4.5–5.5 meters in length, is robotically fabricated using coreless filament winding of flax rovings impregnated with epoxy resin. The components (Figure 14.7) were individually designed based on their structural role, spatial configuration, and fabrication constraints. Their tube-like geometries combine a lattice-like front with high surface curvature to create a stiff surface and reduce the buckling length of the spine, while the back features truss-like layers to add structural depth and transmit wind and snow loads from the polycarbonate roof panels. Each component weighs approximately 100 kg and has been verified through full-scale testing to withstand up to 4 tons of compressive force.

The livMatS Pavilion demonstrates a new paradigm for fiber composites: not as anonymous, high-tech materials, but as performative elements rooted in provenance, material specificity, and ecological intelligence. It reframes composites from synthetically generic to context-sensitive, materially integrated architecture.

Material Scarcity and Hybridity

The building sector consumes over 40% of globally extracted raw materials,[16] underscoring an urgent need to treat all construction resources as finite and scarce, where the ecological value of bio-based materials largely depends on the efficiency of their deployment. Instead of pursuing

FIGURE 14.6 The livMatS Pavilion – A first application of flax fiber composites in architecture.

FIGURE 14.7 Large-scale, load-bearing flax fiber composite component.

an idealized mono-material agenda, hybridity in this context emerges as a pragmatic strategy to increase sustainability through negotiation, mediating between performance, resource use, and local conditions.

In fiber architecture, hybridity refers to integrating structurally interdependent elements made from different materials to achieve performance targets that no single material system could reach alone. They take advantage of the unique properties of each material system – tensile strength, compressive capacity, ductility, lightweight behavior – while mitigating their individual limitations. In this sense, hybridity here does not relate to material mixing but rather to structural and functional interdependence. At the same time, hybrid systems diversify resource use and reduce dependency on high-energy, high-impact materials. This approach also requires a systemic shift in how the construction industry operates. Hybrid systems that rely on renewable resources require transdisciplinary collaboration across agriculture, material science, and construction. As a result, industrial production is increasingly drawn to regions where these natural materials are cultivated and processed, fostering a convergence of regional material cultures and global technological development.

The Hybrid Flax Pavilion (Figure 14.8) exemplifies such a hybrid system, combining robotically wound flax fiber-reinforced polymer components with cross-laminated timber (CLT). Designed as the roof structure for a permanent exhibition building, the roof's 20 hybrid beam elements (Figure 14.9) alternate with 24 standard CLT panels, covering 380 square meters without interior columns. The synergy between flax's tensile capacity and timber's compressive strength enables long-span, lightweight components that satisfy snow load regulations without material excess.[17]

FIGURE 14.8 The Hybrid Flax Pavilion, Landesgartenschau Wangen im Allgäu, 2024.

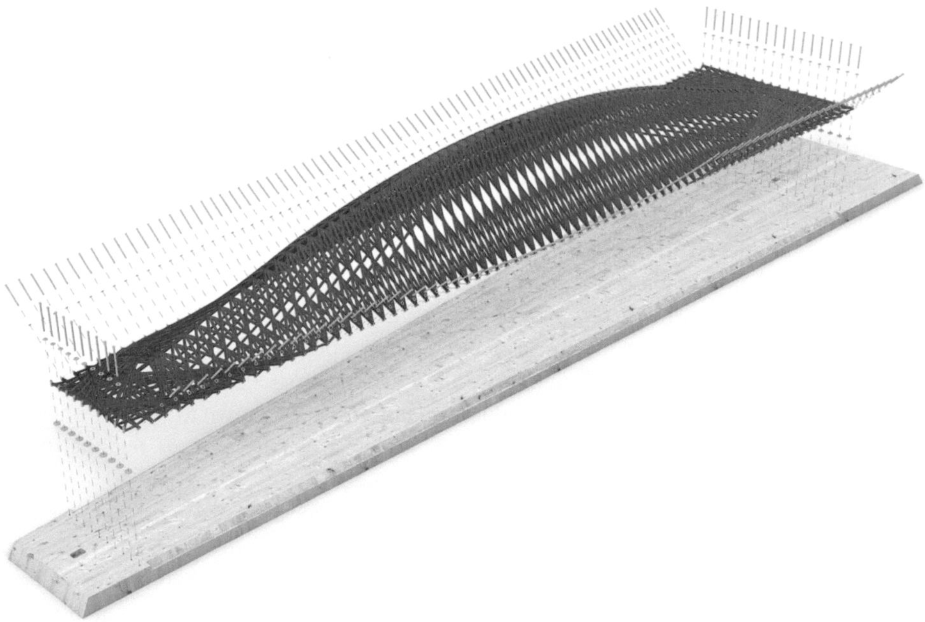

FIGURE 14.9 Hybrid beam elements, combining CLT plates with flax fiber composite.

FIGURE 14.10 Fabrication of hybrid beam elements using standard filament winding equipment.

All fiber components were produced by a local composite manufacturer using standard manufacturing infrastructure by expanding the existing machine logic to increase its capabilities to fabricate fibrous building elements without the need for significant additional investment in new equipment (Figure 14.10). The fiber elements were shipped just in time to a local timber fabricator for final assembly and eventually delivered to the construction site. This tightly coordinated supply chain minimized storage, transport emissions, and waste while enabling rapid on-site installation. Compared to a solid 260 mm-thick CLT plate required to meet equivalent performance benchmarks, the hybrid system uses only 120 mm-thick CLT plates, reducing timber volume by 53.8% and structural weight by 41.5%.[18] The mechanically connected, modular elements can be dismantled, reused, or adapted – an important departure from linear construction logics and a move toward circularity.

Advancing Bio-Based Composites in Construction

Looking ahead, bio-based fiber composites offer a compelling path for transitioning toward more sustainable and circular material cultures. Their potential lies not only in replacing conventional petrochemical composites but in enabling different design, construction, and reuse logics, based on adaptability, efficiency, and regional resilience. The future viability of natural composites such as flax will depend on the development of localized value chains, from cultivation to processing to manufacturing. A systemic evaluation of these opportunities, combining expert insight with economic and environmental modeling, revealed promising trajectories such as substituting slowly regenerating materials like structural timber with bio-based fibers to alleviate pressures on forests while stimulating new forms of rural industry and material supply.

The transition to fully bio-based composites is not without technical challenges. Bio-resins, while promising, currently still exhibit processing difficulties and lower mechanical performance than petrochemical alternatives. However, as resin formulations evolve and manufacturing equipment adapts, the path to high-performance, fully bio-based composite systems becomes more tangible. These advancements will be crucial in closing the gap between ecological ideals and practical application.

Beyond the materials themselves, the broader outlook for fiber composites is closely tied to how we attempt to address the prevalent challenges in construction. With more than 65% of the projected 2060 building stock in OECD countries already in place,[19] the most impactful application of bio-based composite building systems may well be in adaptive reuse. Lightweight, high-strength, and digitally prefabricated composite building elements can support building conversion and extension, offering new life to existing structures without the resource cost of demolition and reconstruction.

For widespread adoption, however, the architectural use of fiber composites must evolve from an experimental niche to a widely accessible method of crafting climate-resilient, resource-sensitive, and spatially adaptive environments. This includes developing strategies for modular assembly, disassembly, and reconfiguration, enabled by near-site prefabrication and compatibility with reused building elements, but also requires overcoming regulatory barriers that currently demand elaborate certifications for each individual project.[20] The future of fiber composites in construction ultimately depends on reconciling technical performance with economic viability and regulatory acceptance, where success will be measured not by the purity of bio-based solutions but by their capacity to transform how buildings adapt, endure, and contribute to a circular material economy.

Acknowledgements

The authors would like to express their gratitude to their fellow researchers at the ICD, ITKE, and the Cluster of Excellence IntCDC, as well as to the students of the ITECH Master's Program, for their invaluable contributions to the presented research demonstrators. We further thank our project partners – FibR GmbH, Bundesgartenschau Heilbronn 2019 GmbH, La Biennale di Venezia 2021, the Cluster of Excellence livMatS at the University of Freiburg, HA-CO Carbon GmbH, Sterk Abbundzentrum GmbH, and the Landesgartenschau Wangen im Allgäu 2024 – for their collaboration.

This research was supported by the Deutsche Forschungsgemeinschaft (DFG, German Research Foundation) under Germany's Excellence Strategy – EXC 2120/1 – 390831618.

Notes

1 Elmar Witten et al., "Composites Market Report 2018: Market Developments, Trends, Outlooks and Challenges," *Carbon Composites*, 2018.

2 Elke Genzel and Pamela Voigt, *Kunststoffbauten: Teil 1. Die Pioniere* (Verlag der Bauhaus-Universität Weimar, 2005).

3 Chiara Bedon, "Review on the Use of FRP Composites for Facades and Building Skins," *American Journal of Engineering and Applied Sciences* 9, no. 3 (2016), https://doi.org/10.3844/ajeassp.2016.713.723.

4 Jan Knippers et al., *Construction Manual for Polymers + Membranes: Materials, Semi-Finished Products, Form-Finding, Design,* Edition Detail (Birkhäuser Verlag, 2011).

5 Jan Knippers and Thomas Speck, "Design and Construction Principles in Nature and Architecture," *Bioinspiration & Biomimetics* 7, no. 1 (2012), https://doi.org/10.1088/1748-3182/7/1/015002.

6 Serban Bodea et al., "Robotic Coreless Filament Winding for Hyperboloid Tubular Composite Components in Construction," *Automation in Construction* 126 (2021), https://doi.org/10.1016/j.autcon.2021.103649.

7 Christoph Zechmeister et al., "Design for Long-Span Core-Less Wound, Structural Composite Building Elements." *Impact: Design With All Senses Proceedings of the Design Modelling Symposium 2019*, 2020, https://doi.org/10.1007/978-3-030-29829-6_32.

8 Christoph Zechmeister et al., "Concurrent, Computational Design and Modelling of Structural, Coreless-Wound Building Components," *Automation in Construction* 151 (2023), https://doi.org/10.1016/j.autcon.2023.104889.

9 Niccolo Dambrosio et al., "Towards an Architectural Application of Novel Fiber Composite Building Systems – the BUGA Fibre Pavilion," *in ACADIA – Ubiquity and Autonomy [Proceedings of the ACADIA Conference 2019]. The University of Texas, Austin*, 2019.

10 Niccolo Dambrosio et al., "Design and Development of an FRP-Timber Hybrid Building System for Multi-Story Applications in Architecture: Maison Fibre," *In ACADIA – Realignments: Toward Critical Computation [Proceedings of the ACADIA Conference 2021]*, 2021.

11 Pascal Mindermann et al., "Investigation of the Fabrication Suitability, Structural Performance, and Sustainability of Natural Fibers in Coreless Filament Winding," *Materials* 15, no. 9 (2022). https://doi.org/10.3390/ma15093260.

12 Ryszard M. Kozlowski et al., "Bast Fibres: Flax," in *Handbook of Natural Fibres* (Elsevier, 2012).

13 Menghe Miao and Niall Finn, "Conversion of Natural Fibres into Structural Composites," *J. Text. Eng.* 54, no. 6 (2008), https://doi.org/10.4188/jte.54.165.

14 Christoph Zechmeister et al., "Extension of Computational Co-Design Methods for Modular, Prefabricated Composite Building Components Using Bio-Based Material Systems," *Sustainability* 15, no. 16 (2023), https://doi.org/10.3390/su151612189.

15 Niccolo Dambrosio et al., "Livmats Pavilion: Design and Development of a Novel Building System Based on Natural Fibres Coreless-Wound Structural Components for Applications in Architecture," *Journal of the International Association for Shell and Spatial Structures* 65, no. 2 (2024), https://doi.org/10.20898/j.iass.2024.009.

16 Jordi Castellano et al., "Integrated System Approach to Evaluate Social, Environmental and Economics Impacts of Buildings for Users of Housings," *Energy and Buildings* 123 (2016), https://doi.org/10.1016/j.enbuild.2016.04.046.
17 Christoph Zechmeister et al., "Design and Development of Natural Fiber-Timber Hybrid Beam Elements for Sustainable Construction," *Architectural Intelligence* 4, no. 1 (2025), https://doi.org/10.1007/s44223-025-00083-6.
18 Monika Göbel et al., "The Hybrid Flax Pavilion: Towards Minimally Invasive Bio-Based Architecture," *IOP Conference Series: Earth and Environmental Science* 1554, no. 1 (2025), https://doi.org/10.1088/1755-1315/1554/1/012095.
19 United Nations Environment Programme, "Towards Zero-Emission Efficient and Resilient Buildings: Global Status Report 2016," (2016).
20 Marta Gil Pérez et al., "Structural Design Assisted by Testing for Modular Coreless Filament-Wound Composites: The BUGA Fibre Pavilion," *Constr. Build. Mater.* 301 (2021), https://doi.org/10.1016/j.conbuildmat.2021.124303.

References

Bedon, Chiara. "Review on the Use of FRP Composites for Facades and Building Skins." *American Journal of Engineering and Applied Sciences* 9, no. 3 (2016): 713–23. https://doi.org/10.3844/ajeassp.2016.713.723.

Bodea, Serban, Christoph Zechmeister, Niccolo Dambrosio, Moritz Dörstelmann, and Achim Menges. "Robotic Coreless Filament Winding for Hyperboloid Tubular Composite Components in Construction." *Automation in Construction* 126 (2021): 103649. https://doi.org/10.1016/j.autcon.2021.103649.

Castellano, Jordi, Albert Ribera, and Joaquim Ciurana. "Integrated System Approach to Evaluate Social, Environmental and Economics Impacts of Buildings for Users of Housings." *Energy and Buildings* 123 (2016): 106–18. https://doi.org/10.1016/j.enbuild.2016.04.046.

Dambrosio, Niccolo, Christoph Zechmeister, Serban Bodea, Valentin Koslowski, Marta Gil Pérez, Bas Rongen, Jan Knippers, and Achim Menges. "Towards an Architectural Application of Novel Fiber Composite Building Systems – The BUGA Fibre Pavilion." In *ACADIA – Ubiquity and Autonomy [Proceedings of the ACADIA Conference 2019]*. ACADIA (Association for Computer Aided Design in Architecture).The University of Texas, Austin, 2019, 234–43.

Dambrosio, Niccolo, Christoph Zechmeister, Rebeca Duque Estrada et al. "Design and Development of an FRP-Timber Hybrid Building System for Multi-Story Applications in Architecture: Maison Fibre." In *ACADIA – Realignments: Toward Critical Computation [Proceedings of the ACADIA Conference 2021]*, 2021.

Dambrosio, Niccolo, Christoph Zechmeister, Marta Gil Pérez et al. "Livmats Pavilion: Design and Development of a Novel Building System Based on Natural Fibres Coreless-Wound Structural Components for Applications in Architecture." *Journal of the International Association for Shell and Spatial Structures* 65, no. 2 (2024): 133–45. https://doi.org/10.20898/j.iass.2024.009.

Genzel, Elke, and Pamela Voigt. *Kunststoffbauten: Teil 1. Die Pioniere*. Verlag der Bauhaus-Universität Weimar, 2005.

Gil Pérez, Marta, Bas Rongen, Valentin Koslowski, and Jan Knippers. "Structural Design Assisted by Testing for Modular Coreless Filament-Wound Composites: The BUGA Fibre Pavilion." *Construction and Building Materials* 301 (2021): 124303. https://doi.org/10.1016/j.conbuildmat.2021.124303.

Göbel, Monika, Christoph Zechmeister, Rebeca Duque Estrada, Jens Weißert, Harrison Hildebrandt, Alina Rossa, Christoph Schlopschnat, Daniel Christopher Bozó, Pascal Mindermann, Jan Knippers, Götz T. Gresser, Franziska Schünemann, Philip Leistner and Achim Menges. "The Hybrid Flax Pavilion: Towards Minimally Invasive Bio-Based Architecture." In *IOP Conference Series: Earth and Environmental Science* 1554 (2025): 012095. https://doi.org/10.1088/1755-1315/1554/1/012095

Knippers, Jan, Jan Cremers, Markus Gabler, and Julian Lienhard. *Construction Manual for Polymers + Membranes: Materials, Semi-Finished Products, Form-Finding, Design*. Edition Detail. Birkhäuser Verlag, 2011.

Knippers, Jan, and Thomas Speck. “Design and Construction Principles in Nature and Architecture.” *Bioinspiration & Biomimetics* 7, no. 1 (2012): 15002. https://doi.org/10.1088/1748-3182/7/1/015002.

Kozlowski, Ryszard M., Maria Mackiewicz-Talarczyk, and Anwar M. Allam. “Bast Fibres: Flax.” In *Handbook of Natural Fibres*, edited by Ryszard M. Kozlowski. Elsevier, 2012, 56–113.

Miao, Menghe, and Niall Finn. “Conversion of Natural Fibres into Structural Composites.” *Journal of Textile Engineering* 54, no. 6 (2008): 165–77. https://doi.org/10.4188/jte.54.165.

Mindermann, Pascal, Marta Gil Pérez, Jan Knippers, and Götz T. Gresser. “Investigation of the Fabrication Suitability, Structural Performance, and Sustainability of Natural Fibers in Coreless Filament Winding.” *Materials* 15, no. 9 (2022). https://doi.org/10.3390/ma15093260.

United Nations Environment Programme. “Towards Zero-Emission Efficient and Resilient Buildings: Global Status Report 2016.” 2016.

Witten, Elmar, Volker Mathes, Michael Sauer, and Michael Kühnel. “Composites Market Report 2018: Market Developments, Trends, Outlooks and Challenges.” *Carbon Composites* (2018): 1–59.

Zechmeister, Christoph, Serban Bodea, Niccolo Dambrosio, and Achim Menges. “Design for Long-Span Core-Less Wound, Structural Composite Building Elements.” *Impact: Design with All Senses Proceedings of the Design Modelling Symposium 2019.* Springer 2020, 401–415. https://doi.org/10.1007/978-3-030-29829-6_32.

Zechmeister, Christoph, Harrison Hildebrandt, Rebeca Duque Estrada et al. “Design and Development of Natural Fiber-Timber Hybrid Beam Elements for Sustainable Construction.” *Architectural Intelligence* 4, no. 1 (2025). https://doi.org/10.1007/s44223-025-00083-6.

Zechmeister, Christoph, Marta Gil Pérez, Niccolo Dambrosio, Jan Knippers, and Achim Menges. “Extension of Computational Co-Design Methods for Modular, Prefabricated Composite Building Components Using Bio-Based Material Systems.” *Sustainability* 15, no. 16 (2023): 12189. https://doi.org/10.3390/su151612189.

Zechmeister, Christoph, Marta Gil Pérez, Jan Knippers, and Achim Menges. “Concurrent, Computational Design and Modelling of Structural, Coreless-Wound Building Components.” *Automation in Construction* 151 (2023): 104889. https://doi.org/10.1016/j.autcon.2023.104889.

15

MALLEABILITY AS A DRIVER FOR DESIGN

Activating Alternative Resource Flows with 3D-Printed Biopolymer Composites

Mette Ramsgaard Thomsen, Gabriella Rossi, Carl Eppinger, and Paul Nicholas

Introduction

In Faircloth's seminal work *Plastics Now*—which unfolds like a detective narrative unravelling the concealed histories of plastics in architecture—the author introduces a satellite image of Baytown, Texas, home to both ExxonMobil and the Bayer MaterialScience Industrial Park.[1] Through this image, Faircloth challenges the reader to comprehend the immense scale of plastics production. Expanding our focus from the isolated properties, performances, and aesthetics of plastics, she directs our attention to the territorial consequences of its production. These thousand-hectare "multi-story chemistry sets" and the sites of resource extraction for both fossil-fuel-based and bio-based plastics occupy landscape-scale spaces. By broadening our perspective, she interrogates how production processes incorporate territorial expanses that, in turn, engender architectural consequences.

This chapter asks similar questions of our material practices and their interconnectedness with their local terroir of provenance. With focus on bio-based plastics, and extrudable biopolymer composites in particular, we ask how these highly versatile and responsive materials can be incorporated into architectural building practice. Biopolymer composites are fundamentally different to their fossil fuel counterparts. While preceding their invention they retain an environmental responsiveness that witnesses their ecological belonging. The chapter asks how we can conceive the application of biopolymer composites in architecture through a concept of malleability that stretches through the scales of intervention. From their material composition and recipe logic, through their fabrication procedures, deployment in our built environment and finally to the way that they occupy their own territorial footprints, we ask how ideas of malleability can structure the way we work with what is becoming a renewed interest in an old class of materials. With point of departure in the demonstrator project Rhizerial, we develop this inter-scalar thinking through a located case study, in order to outline the ways in which this multiplicity of *malleabilities* can be enfolded through a contemporary architectural practice with biopolymer composites.

DOI: 10.4324/9781003613589-18

Returns to Biopolymers Composites in Construction

Biopolymer composites are not new to construction. Preceding the fossil fuel counterparts, they represent a significant yet often under-acknowledged chapter in material science. The following traces the historical trajectory of biopolymer composites, situating their early development within 19th-century material innovation and examining their overlooked significance in shaping industrial economies. It explores how these early materials—derived from organic residues and often tied to colonial supply chains—prefigured today's sustainability challenges by embedding environmental and geopolitical consequences into their production. Revisiting this history allows for a critical reappraisal of contemporary biopolymer composites in architecture, not only as sustainable alternatives but as agents entangled in complex agro-ecological and socio-material systems.

Early biopolymers were predominantly derived from naturally occurring substances such as starch, cellulose, casein, shellac, and ebonite. These materials emerged as pioneering alternatives to conventional materials.[2] The history of biopolymer composites reads like a primer for the material innovation of the 19th century, where for example, materials like Bois Durci, made from hardwood sawdust, carbon black, metal particles, and blood or egg, were used in design products like mirrors.[3] Casein, a curdled milk protein obtained through acidification or enzymatic action, was used to make buttons and jewellery, and Celluloid was famously used in the production of billiard balls[4] and later into a host of products, coatings, and as a film base. Then, like now, biopolymer composites were developed in response to pressure on resources. Celluloid's narrative of making is directly contextualized within the contemporary American billiard craze[5] resulting from a rise of cost ivory and a threat to elephants.[6] Made from various cellulose sources including cotton, ramie, flax, jute, and wood pulp,[7] much of its production happened through residues from the cotton mill industries. At the same time, it relied on natural camphor grown in Asia, namely Taiwan, and the Gulf of Mexico, namely Florida,[8] where increasing demand lead to new enforced camphor trade treaties, resulting in a collapse of its value and therefore availability for the emerging plastics industry.[9] In this way, this particular track of material invention draws new agro-geographical boundaries that extend deep into colonialist histories drawing in territories far removed from their application and use.[10]

These early biopolymer composites made from experimental, malleable, recipes, alongside others such as natural compounds, rubber and silk, played a central role in industrial advancement during the 19th and early 20th centuries representing significant economies. With the rise of synthetic plastics, such as Bakelite and later polyesters, the use of biopolymer composites diminished as they were seen as less durable and more expensive to produce at scale. Yet, their synthetic counterparts share the sense of this remote geography. Fossil fuel-based polymers have moved its implicated resource territories underground, to become part of what Frederique Aiit-Touati calls "host objects, the inert elements which alter the composition of the living soil and constitute the real soil of anthropized territories."[11] Just as early on the ecological consequences of biopolymer composite production were rendered invisible[12] by their remoteness, so is contemporary fossil fuel-based polymer production, its waste and pollution hidden from view.

The contemporary desire to return to biogenic materials in design and architecture as sustainable alternatives to a building practice out of balance has positioned biopolymer composites at the forefront of a new material revolution.[13] Renewable, with a low carbon footprint and an ability to biodegrade, biopolymer composites are part of a promise to upend the extractive practices of finite materials. However, for biopolymer composites to fulfil their sustainable potential, it is crucial to rethink the agro-geographies that have historically accompanied their production. This involves addressing the new pressures biopolymer composite production places on socioecological networks and the biodiversity they sustain in their respective local terroirs. Placing biopolymers within a critical thinking of circular design foregrounds the ability to explore the role of locality and the way that residual waste streams within the bio-domain can be strategically deployed while remaining sensitive to the dependencies that they entail.[14] Using the thinking of *Reciprocal Landscapes*, as framed by Jane Mah Hotton,[15] this new era of biopolymer composites and its application in architecture must engage with its surrounding ecosystem in a relational way, considering not only the ecological benefits of materials but also the broader systems of land, labor, and resources they depend on.

Malleability as an Inter-Scale Material Perspective on Circularity

In this chapter, we link the principles of circular design for architecture to practices of malleability around the design and fabrication with local recipes of biopolymer composites. Malleability as a concept is here understood on two levels, the first of which is physical. Malleability and its counterpart plasticity are concepts used in material science when defining properties of polymers and their composites. Plasticity refers to a material's ability to undergo permanent deformation when subjected to stress, meaning it retains a new shape after the force is removed.[16] In architecture, plasticity often focuses on the material's capacity to be shaped, molded, or formed into complex geometries. It implies a lasting change in the material's structure, often leveraged in design to create highly expressive, sculptural forms. In difference, malleability focuses on the ability of a material to be repeatedly reshaped or altered without permanently losing its inherent properties.[17] Unlike plasticity, which implies a permanent transformation, malleability highlights the dynamic and continuous modification of materials across multiple scales.

The second level of understanding is more strategic addressing resource provenance and the way resource procurement is part of localized agricultural practices of growing and harvesting. Here, malleability centers on adaptability and replaceability, articulating how a recipe, or material composition, can be adjusted, modified, or replaced in response to shifts in availability, or changes to fabrication constraints. This includes the strategic substitution of key ingredients to accommodate resource availability, the steering of fabrication processes, and the material's capacity to adapt throughout its lifecycle. At both levels, malleability becomes a critical concept for circular design providing a vocabulary for how to think the relationality of biogenic material systems, and biopolymer composite in particular, to their associated geographies to remain flexible, responsive, and resilient.

Rhizerial—An Exploration of Malleable Material Models through Functionally Graded Structures

In our work, malleability serves as a key concept for understanding the inter-scalar behavior of biopolymer composite material systems, both on a physical, as well as a strategic level as outlined in the section above. On a physical level, biopolymer composites are malleable materials.[18] Their mechanical properties allow them to be flexible under load, while their chemical properties make them both hygroscopic and thermally responsive to their environments. The elements made of biopolymer composites themselves can be materially graded, combining the properties of a gradient of shifting materials. On a strategic level, biopolymer composites are malleable with regards to their recipe composition, the mixing of their constituents, and their fabrication processes. At these distinct interconnected scales, the concept of malleability enables a foregrounding of flexibility and responsiveness, allowing resources to be selected, designed, and deployed dynamically based not only on their performance, but also on availability and environmental impact, while also ensuring their reintegration into material cycles through reclamation and biodegradation. This perspective positions malleability as a strategic framework for designing adaptable and regenerative material systems, such as biopolymer composites.

These interconnected scales of design are demonstrated in the 3D-printed Rhizerial installation (Figure 15.1) developed for the 'In the Making – Artistic Methods in Architecture and Design' exhibition, Royal Danish Academy in 2024. Here, material grading is used as a design driver to instrumentalize and showcase malleability at the different scales of the biopolymer composite. In the installation, the 14 vaulted panels are designed to be geometrically porous, with volumetric lattice structures allowing light to penetrate through, and materially graded, with mixes of biopolymer composite recipes. The panels are robotically 3D printed using a bespoke grading system. The installation leverages a series of computational infrastructures developed to enable

FIGURE 15.1 Rhizerial installation at the Royal Danish Academy. The 3D-printed vaults are continuously graded showcasing the potentials of waste-stream biopolymer composites in architecture. © Anders Ingvartsen and CITA 2024.

this inherent malleability, which we detail in the upcoming section. Firstly, the *Recipe Explorer*, a Machine-Learning based mapping interface allowing a distribution of resource inputs in relation to their performances that can be strategically navigated in response to availability and performance. Secondly, the tool pathing workflow and dual-pump 3D-printing hardware that allows us to translate a designed material shift intent into physical printed beads and finally, the tracking of environmental response and reforming processes undertaken in response to its dynamic environmental context.

Unfolding the Scales of Malleability through Design to Fabrication Workflows

From Design Spaces to Recipe Spaces

In conventional practice, architects are primarily engaged in geometrical modeling within a determined design space, to be understood as the space of variation of the parameters that drive the design. Material choices are often secondary to formal considerations. Rediscovering biopolymer composites as a valuable class of architectural materials forces us to rethink our material practice within the design process. Materials are no longer a categorical choice defined by standardized industrial products, but rather a designable range of mixtures and compositions, which we define as the *Recipe Space.*[19] Exploring the recipe space in search of material mixtures becomes a central part of the design process. The architect therefore assumes a role akin to that of a material scientist, actively shaping the composition, and resulting behavior of novel and alternative composite material systems.[20] This shift carries profound implications for digital modeling, as the need to represent heterogeneous and dynamic materials, such as 3D-printed biopolymer composites, exceeding the capacities of conventional architectural modeling infrastructures.

In Rhizerial, the *Recipe Space* is that of a collagen-based biopolymer composite (Figure 15.2). The collagen glue binder itself is a by-product of the treatment of bones and tissues from the meat industry. Water, glycerol, and collagen are heated and mixed to create a binder matrix, which is subsequently bulked and reinforced with waste-stream sourced fibers, which have been sourced locally in Denmark. In Rhizerial, four cellulose-based fibers are used: seagrass sourced from the Danish coast south of Copenhagen, cotton fibers stemming from recycled bedsheets from the hospitality industry, sawdust and bark which are by-products of the Danish timber industry, specifically consisting of Beech trees.[21] The process results in a thermoplastic material, which is heated to 65°C in order to attain a viscosity that can be robotically extruded. Once cool, the material hardens and retains its shape. 3D printing is followed by a curing process which can take up to two weeks depending on the environmental factors. During this process, the material cures through the evaporation of the water content. The hygroscopic behavior of the material recipe lies at the core of its physical malleability: the material retains an ability to absorb environmental humidity. When exposed to fluctuating environmental conditions it responsively shapeshifts as water content dynamically fluctuates. This property can also be actively triggered in post-production reforming by spraying the panels with water and laying them in forms to either reshape or correct their geometry. Finally, the same property also makes the material fully circular as it can be rehydrated, reheated, and reprinted, with no detriments to its mechanical properties.[22] This material malleability affords the material with an inherent circularity which is different to the capacities of fossil fuel-based counterparts.[23]

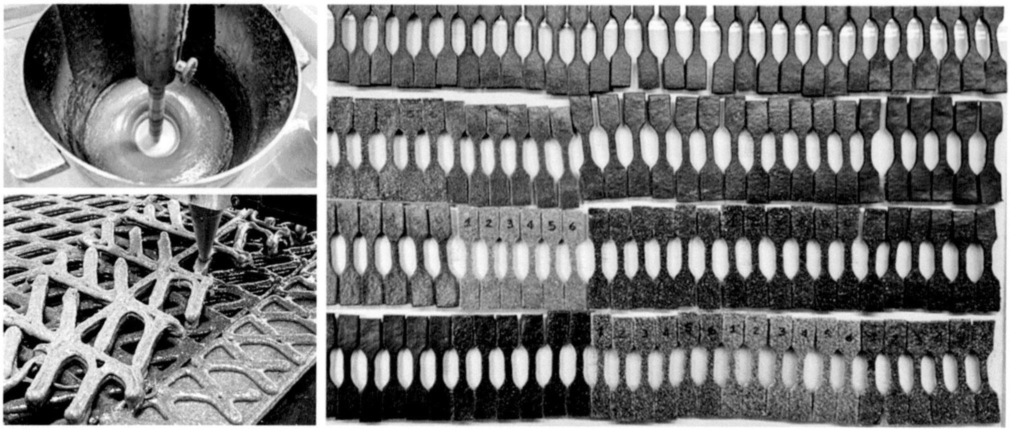

FIGURE 15.2 Recipe variations of Collagen glue biopolymer composite. (top left) Matrix preparation: Bark fibers being dispersed into the binder matrix. (bottom left) Sequential printing with different fibers: yellow wood flour, dark brown seagrass and light blue cotton. (right) Dog bone material samples for prepared for characterization through destructive mechanical testing. © CITA 2024.

Mapping of Material Resources in the Recipe Explorer

The material design logics introduced by the usage of biopolymer composite systems point towards an alternative design-driven computational modeling paradigm. This change questions our logic of resource use and performance-driven design. Traditionally, architectural material strategies have been shaped by a framework of scarcity, emphasizing efficiency-driven methodologies such as material minimization, topological optimization, and the pursuit of high-performance composites. In Rhizerial, we explore an alternate paradigm that embraces malleability, adaptability, redundancy, and thickness, where substitution and reconfiguration become intrinsic to the design process, and where material properties carry an associated tolerance. Rather than thinking about material composition as a singular formulation, or an optimized Pareto-front of formulations, we conceptualize the *Recipe Space*, as an explorable design space, where formulations of the biopolymer composites and negotiation between their properties. Material properties are not limited to mechanical or thermal performances, but we can integrate different ideas of performance such as locality, carbon footprint, resource availability, and aesthetics, and navigable performance landscapes which can be negotiated in the design process. In Rhizerial, the *Recipe Explorer* (Figure 15.3) is a concrete materialization of material malleability at a conceptual level. It allows to represent materials with their associated dynamic properties and tolerances, and in so doing enable validation of matrial choices, evaluations, and performance range compromises to adapt for availability and locality, and aesthetic expression beyond optimizations and standardization.

The *Recipe Explorer* is a computational infrastructure which we have developed to exercise malleability with respect to 3D-printing biopolymer composites, physically and conceptually. The engine uses unsupervised Machine Learning to produce an associative map of the mixture variations within our *Recipe Space*. For Rhizerial, the included material formulations are with fiber concentrations ranging from 0 to 30 Wt%, with a maximum of two fibers present in one

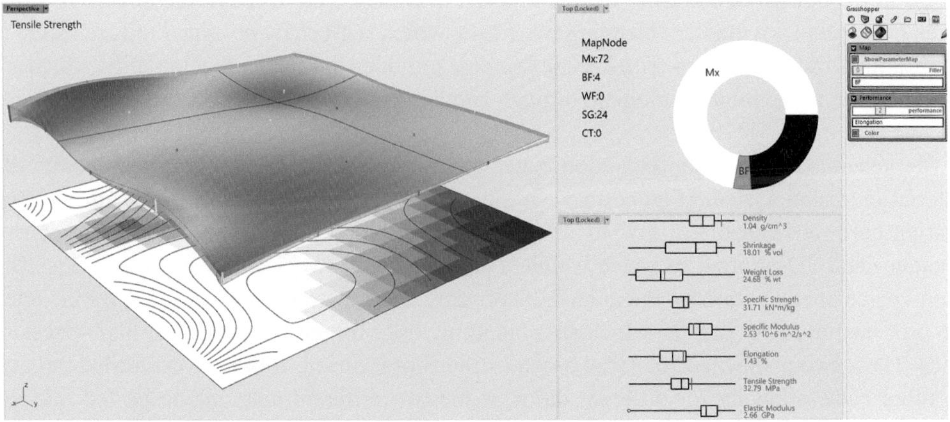

FIGURE 15.3 The recipe explorer UI integrated within the CAD software. Here the user is hovering over recipe mixing 72% collagen glue matrix, with 24% seagrass and 4% Bark fibers. The associated predicted performances as well as the bounds of the performance landscape are shown to the user to facilitate material-design decision making, supported by a a seamless data flow to the downstream processes. © Gabriella Rossi 2024.

mixture at a time. This ensures a printable viscosity under 300 PaS at 65°C and compatible with our printing hardware. Overlaid on top of the formulation map, are the performance landscapes, which are interpolated thanks to a Machine Learning model trained on a curated dataset of physical samples.[24] The *Recipe Explorer* is presented to the designer via an interactive User Interface directly within Rhino/Grasshopper CAD software, where they can explore the trained map and landscapes, and see the recipe composition, its different performances, statistics of the entire *Recipe Space*, as well as uncertainties and tolerances related to the data, irrespective of whether the recipe has been physically tested or not. New layers of performance landscapes can also be subsequently added to the map at later data-gathering stages. Furthermore, the mapping system is inherently applicable to other biopolymer composite formulations, and given the availability of the data to be tuned on, would be able to incorporate a swap of fibers, due to changes in sourcing or seasonal availability, therefore leveraging material malleability.

This computational infrastructure allows for simultaneous coupling of material composition and material properties, which unlocks aspects of design capacity and decision-making which were not previously possible. This method also shifts our understanding of material modeling in architecture, from mono-material assignments onto geometry, and towards more complex multi-material understandings which leverage material malleability, specifically for biopolymer 3D printing. By defining paths of mixture variation along the maps, grading of 3D printed biopolymer elements becomes possible, where we can steer aesthetic and performative variation while remaining aware of the constituent waste-stream fiber resource footprint of the structure.

From Material Specification to Fabrication Data

The Recipe Explorer was instrumental in the development of *Rhizerial*. Once the intricate 3D geometry of the vaulted panels was generated, then came a process of designing with *Recipe Space*. The material design strategy follows a circular pattern at the top of the panels and a linear pattern at the bottom of the panels, which morph into each other through the thickness of the panel. Each panel is a gradient blending two fibers, which is given by printing limits reflected in the variations of the mixtures. The *Recipe Explorer* provided the steps of recipe shifts which are then allocated to the printing toolpath through another Machine-Learning based computational pipeline (Figure 15.4).

The generated fabrication data coordinates the motion of the robot and turntable with the material extrusion system (Figure 15.5). A significant part of developing *Rhizerial* involved creating hardware capable of graded 3D-printing with biopolymers. The robotic arm carries a custom-built 3D-printing extruder, which is equipped with two separate material inlets. The central piece of equipment for the graded 3D printing is the material feed. For this, we included two progressive cavity pumps, which are commonly used for slurries and other high-viscosity fluids. These two pumps feed material based on the chosen speed, which we controlled through the robot code, allowing for different ratios of the two different materials to be fed into the extruder. By translating each recipe into specific pump speed ratios, we achieved the desired fiber blend entering the extruder. Once in the extruder, the materials are mixed with a continuous auger, which is also designed to allow for a consistent extrusion flow. The material needs to be continuously heated throughout the different stages of the printing process, as the biopolymer's viscosity directly relates to its temperature.

Through the making of Rhizerial, we have shown how to instrumentalize malleability, both at a physical level, through the development of a biopolymer composite material which retains characteristics of physical malleability through dynamic material behavior and capacity for manufacturing graded architectural elements, and at the strategic level, through the development of a computational infrastructure which couples different understandings of material properties and performance, and integrates them within the design logics and processes.

Conclusion: Malleability as a Strategy for Resilient Design

In the book *Down to Earth*, Bruno Latour rhetorically asks how to respond to climate crises: "Go backwards? Relearn the old recipes? Take a new look at the age-old wisdom? ... Yes, of course, but without lulling ourselves into with illusions: for them, too, there is no precedent."[25] As an

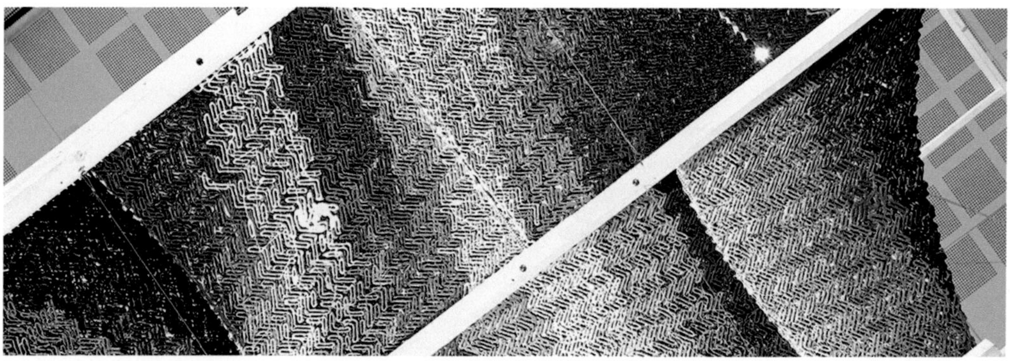

FIGURE 15.4 View from below Rhizerial showcasing the printed material gradient across the vaulted panels (top) blending wood flour to cotton (bottom) wood flour to seagrass. © Anders Ingvartsen and CITA 2024.

FIGURE 15.5 Robotic fabrication setup for Rhizerial. The turntable holds a curved printing jig fitted with a metal mesh. This allows the robotic arm to remain in a stable printing angle throughout the process. The custom-built extruder takes into two matrial feeds from the dual-pump, in this case collagen glue mixed with cotton fibers (blue) and with sawdust (yellow). © CITA 2024.

FIGURE 15.6 The vaulted panels from above. © Anders Ingvartsen and CITA 2024.

FIGURE 15.7 Grading patterns along the vault lines. © Anders Ingvartsen and CITA 2024.

examination of biopolymer composite material in architecture, Rhizerial could be understood in this context. However, rather than return to cemented knowledge rocked by an unprecedented crisis, Rhizerial is more a return to the forgotten starting points that launched a new era of material thinking. Early biopolymers, in all their versatility, emerged from a 'chemistry of gunks'[26] through a wealth of experimentation with recipes and forming tools. They were abandoned because they were deemed unstable, too environmentally responsive and too prone to degradation. But in the context of today's climate crisis, could these same characteristics—malleability, adaptability and circularity—be reimagined as assets rather than deficiencies?

Malleability, in this sense, becomes a strategy of resilience. At the material scale, it allows for reprocessing, adaptation, and reclamation. At the design scale, it challenges conventional architectural specification, foregrounding variability and responsiveness as integral to both performance and aesthetics. And at the territorial scale, it prompts a rethinking of resource geographies, ensuring that material systems remain attuned to ecological flux, seasonal cycles, and local conditions. A central aspect of this is the creation of new methods that allow designers to become aware of the consequences of the resource implication of design. This awareness demands tools that engage with the agro-geographies of materials, enabling composition to shift dynamically based on availability and environmental impact. By embedding these considerations into architectural practice, we move away from an illusion of permanence and stability toward an architecture that is adaptive, regenerative, and deeply entangled with its environments. This requires not only new materials but also new design methodologies—computational tools, adaptive fabrication techniques, and circular strategies that both acknowledge but also operate within the shifting material practices of a changing world. *Rhizerial* embodies this ambition, demonstrating how malleability can be strategically mobilized to create an architecture that does not resist change, but works with it—learning, adapting, and evolving in response to the dynamic conditions of the present.

Notes

1 Billie Faircloth, *Plastics Now: On Architecture's Relationship to a Continuously Emerging Material* (Routledge Taylor & Francis Group, 2015), 39.
2 Tim A. Osswald and Sylvana García-Rodríguez, "History of Sustainable Bio-Based Polymers," in *A Handbook of Applied Biopolymer Technology: Synthesis, Degradation and Applications*, ed. Sanjay K. Sharma, 1st, RSC green chemistry 12 (Royal Society of Chemistry, 2011).
3 Osswald and García-Rodríguez, "History of sustainable bio-based polymers," 2.
4 Osswald and García-Rodríguez, "History of sustainable bio-based polymers," 8.
5 Robert Douglas Friedel, *Pioneer Plastic: The Making and Selling of Celluloid* (University of Wisconsin Press Madison, 1983); Tim Palucka, "Artificial Billiard Balls," *MRS Bulletin* 30, no. 8 (2005), https://doi.org/10.1557/mrs2005.186.
6 "When Plastic Saved the Elephants: Pool Balls and the Bakelite Revolution," Paul Shapiro, accessed November 21, 2023, https://paulshapiro.medium.com/when-plastic-saved-the-elephants-pool-balls-and-the-bakelite-revolution-a5d56bb03d37.
7 Julie A. Reilly, "Celluloid Objects: Their Chemistry and Preservation," *Journal of the American Institute for Conservation* 30, no. 2 (1991).
8 Reilly, "Celluloid objects: their chemistry and preservation," 149.
9 Matthew T. Combs, "Camphor, a Plastic History: China, Taiwan, and Celluloid, 1868–1937" (Doctor of Philosophy in History, University of California, Irvine, 2018), accessed September 29, 2025, https://escholarship.org/uc/item/91g9s5tt, 53.
10 Marek Jancovic, "The 'Raw Nature' of Cinema: Rethinking Technological Histories of Film Through Plant and Animal Materials in Early Celluloid Manufacturing," in *NECS 2024 Annual Conference: Emergencies-Media in an Unpredictable World* (2024).
11 Frédérique Aït-Touati et al., *Terra Forma: A Book of Speculative Maps*, trans. Amanda DeMarco (The MIT Press, 2022), 45.
12 Bruno Latour, *Pandora's Hope: Essays on the Reality of Science Studies* (Harvard university press, 1999).
13 Andrea Lipps et al., eds., *Nature: Collaborations in Design* (Cooper Hewitt Smithsonian Design Museum, 2019).
14 Mette Ramsgaard Thomsen et al., "Extending the Circular Design Framework for Bio-Based Materials: Reconsidering Cascading and Agency Through the Case of Biopolymer Composites," in *Design for Rethinking Resources: Proceedings of the UIA World Congress of Architects Copenhagen 2023*, ed. Mette Ramsgaard Thomsen, Carlo Ratti and Martin Tamke, Sustainable Development Goals Series (Springer International Publishing; Imprint Springer, 2024).
15 Jane Elizabeth Hutton, *Reciprocal Landscapes: Stories in Material Movement* (Routledge, 2020).
16 Rodney Hill, *The Mathematical Theory of Plasticity* (Clarendon Press, 2009).
17 Hill, *The Mathematical Theory of Plasticity.*
18 Paul Nicholas et al., "Biopolymer Composites in Circular Design: Malleable Materials for an Instable Architecture," in *Acadia23: Habits of the Anthropocene: Scarcity and Abundance in a Post-Material Economy*, ed. Assia Crawford et al. 2 (2023).
19 Gabriella Rossi, "Designing Small Datasets: ML modeling applications for 3D printed biopolymer composites in architecture" (Royal Danish Academy—Architecture, Design, Conservation, 2024).
20 Mette Ramsgaard Thomsen, "Complex Modelling: Questioning the Infrastructures of Information Modelling," in *Proceedings of the 34th International Conference on Education and Research in Computer Aided Architectural Design in Europe (ECAADe) [Volume 1]*, ed. Aulikki Herneoja, Toni Österlund and Piia Markkanen, eCAADe proceedings (eCAADe, 2016).
21 A detailed description of the material preparation quantities can be found at Arianna Rech et al. (2024).

22 Nicholas et al., "Biopolymer Composites in Circular Design"; Rech et al., "Thermoprocessable Bio-waste-Based Composites from Animal Glue and Lignocellulosic Waste Fibers."
23 Nicholas et al., "Biopolymer Composites in Circular Design."
24 Gabriella Rossi et al., "Predicting the Recipe: Leveraging Prameterless Self-Organzizing Maps (PLSOM) With Geometrical Polynomial Fitting (GPF) To Predict Biopolymer Composites Characterization with Small Datasets," in *The Future of Construction 2023 Symposium* (2023); Rossi, "Designing Small Datasets."
25 Bruno Latour, *Down to Earth: Politics in the New Climatic Regime*, English edition (Polity, 2018), 44.
26 Jeffrey L. Meikle, *American Plastic: A Cultural History* (Rutgers University Press, 1997), 36.

References

Aït-Touati, Frédérique, Alexandra Arènes, and Axelle Grégoire. *Terra Forma: A Book of Speculative Maps*. Translated by Amanda DeMarco. The MIT Press, 2022.

Combs, Matthew Tyler. *Camphor, a Plastic History: China, Taiwan, and Celluloid, 1868-1937*. Doctor of Philosophy in History, University of California, Irvine, 2018. Accessed September 29, 2025. https://escholarship.org/uc/item/91g9s5tt.

Faircloth, Billie. *Plastics Now: On Architecture's Relationship to a Continuously Emerging Material*. Routledge Taylor & Francis Group, 2015.

Friedel, Robert Douglas. *Pioneer Plastic: The Making and Selling of Celluloid*. University of Wisconsin Press Madison, 1983.

Hill, Rodney. *The Mathematical Theory of Plasticity*. Clarendon Press, 2009.

Hutton, Jane Elizabeth. *Reciprocal Landscapes: Stories in Material Movement*. Routledge, 2020.

Jancovic, Marek. "The 'Raw Nature' of Cinema: Rethinking Technological Histories of Film Through Plant and Animal Materials in Early Celluloid Manufacturing." In *NECS 2024 Annual Conference: Emergencies-Media in an Unpredictable World*. 2024. NECS — European Network for Cinema and Media Studies (conference host organization). İzmir University of Economics, İzmir, Turkey.

Latour, Bruno. *Pandora's Hope: Essays on the Reality of Science Studies*. Harvard University Press, 1999.

Latour, Bruno. *Down to Earth: Politics in the New Climatic Regime*. English edition. Polity, 2018.

Lipps, Andrea, Matilda McQuaid, Caitlin Condell, and Gène Bertrand, eds. *Nature: Collaborations in Design*. Cooper Hewitt Smithsonian Design Museum, 2019.

Meikle, Jeffrey L. *American Plastic: A Cultural History*. Rutgers University Press, 1997.

Nicholas, Paul, Ayoub Lharchi, and Martin Tamke et al. "Biopolymer Composites in Circular Design: Malleable Materials for an Instable Architecture." In *Acadia23: Habits of the Anthropocene: Scarcity and Abundance in a Post-Material Economy*. Edited by Assia Crawford et al., 166–73. Association for Computer Aided Design in Architecture (ACADIA), 2023.

Osswald, Tim A., and Sylvana García-Rodríguez. "History of Sustainable Bio-Based Polymers." In *A Handbook of Applied Biopolymer Technology: Synthesis, Degradation and Applications*. Edited by Sanjay K. Sharma. 1st. RSC green chemistry 12. Royal Society of Chemistry, 2011.

Palucka, Tim. "Artificial Billiard Balls." *MRS Bulletin* 30, no. 8 (2005): 614. https://doi.org/10.1557/mrs2005.186.

Ramsgaard Thomsen, Mette. "Complex Modelling: Questioning the Infrastructures of Information Modelling." In *Proceedings of the 34th International Conference on Education and Research in Computer Aided Architectural Design in Europe (ECAADe) [Volume 1]*. Edited by Aulikki Herneoja, Toni Österlund and Piia Markkanen, 33–42. eCAADe proceedings. eCAADe, 2016.

Ramsgaard Thomsen, Mette, Gabriella Rossi, Anders Egede Daugaard, Arianna Rech, and Paul Nicholas. "Extending the Circular Design Framework for Bio-Based Materials: Reconsidering Cascading and Agency Through the Case of Biopolymer Composites." In *Design for Rethinking Resources: Proceedings of the UIA World Congress of Architects Copenhagen 2023*. Edited by Mette Ramsgaard Thomsen, Carlo Ratti and Martin Tamke, 635–46. Sustainable Development Goals Series. Springer International Publishing; Imprint Springer, 2024.

Rech, Arianna, Paul Nicholas, Nicolas Møller Jakobsen, Mette Ramsgard Thomsen, and Anders E. Daugaard. "Thermoprocessable Bio-Waste-Based Composites from Animal Glue and Lignocellulosic Waste Fibers." *ACS Sustainable Resource Management* 1, no. 1 (2024): 54–64. https://doi.org/10.1021/acssusresmgt.3c00032.

Reilly, Julie A. "Celluloid Objects: Their Chemistry and Preservation." *Journal of the American Institute for Conservation* 30, no. 2 (1991): 145–62.

Rossi, Gabriella. *Designing Small Datasets: ML Modeling Applications for 3D Printed Biopolymer Composites in Architecture*. Royal Danish Academy – Architecture, Design, Conservation, 2024.

Rossi, Gabriella, Arianna Rech, John Harding et al. "Predicting the Recipe: Leveraging Prameterless Self-Organzizing Maps (PLSOM) With Geometrical Polynomial Fitting (GPF) To Predict Biopolymer Composites Characterization with Small Datasets." In *The Future of Construction 2023 Symposium.* Technical University of Munich (TUM), Munich, Germany, 2023.

Shapiro, Paul. "When Plastic Saved the Elephants: Pool Balls and the Bakelite Revolution." Accessed November 21, 2023. https://paulshapiro.medium.com/when-plastic-saved-the-elephants-pool-balls-and-the-bakelite-revolution-a5d56bb03d37.

INDEX

Note: Pages indicating figures are in *italics*. Pages indicating tables are in **bold** type.

For Product Safety Concerns and Information please contact our EU representative GPSR@taylorandfrancis.com Taylor & Francis Verlag GmbH, Kaufingerstraße 24, 80331 München, Germany

Batch number: 10722134

Printed by Printforce, the Netherlands